David Cahan
May 1991
Lincoln

The Royal Commission on Historical Manuscripts

Guides to Sources for British History
based on the National Register of Archives

2

The Manuscript Papers of
BRITISH SCIENTISTS
1600 – 1940

London Her Majesty's Stationery Office

First published 1982

HER MAJESTY'S STATIONERY OFFICE

Government Bookshops

49 High Holborn, London WC1V 6HB
13a Castle Street, Edinburgh EH2 3AR
41 The Hayes, Cardiff CF1 1JW
Brazennose Street, Manchester M60 8AS
Southey House, Wine Street, Bristol BS1 2BQ
258 Broad Street Birmingham B1 2HE
80 Chichester Street, Belfast BT1 4JY

Government Publications are also available through booksellers

ISBN 0 11 440122 5

Preface

This Guide is the result of an enquiry initiated by a joint committee of the Commission and the Royal Society on scientific and technological records, set up in 1966 under the chairmanship at first of the late Sir Harold Hartley and afterwards of Professor Nicholas Kurti. The committee's activities formed part of a more widespread movement to encourage and develop public interest in the preservation and study of papers relating to the history of science in which the Commission had already been involved since 1959.

The earlier stages of the research involved were financed by a generous grant from the Goldsmiths' Company and conducted successively by Dr REW Maddison and Mr WJ Craig, with active support throughout from the National Register of Archives. When the joint committee was dissolved in 1976, having fulfilled its intended promotional purposes, the task of completing and revising the guide for press passed entirely to the Commission, where the concluding stages of its preparation have been carried out under the direction of Miss Sonia P Anderson.

When the enquiry began, very little was known about the extent and locations of surviving collections of papers of British men of science. Many of these collections lay unnoticed and unlisted in the places where they had been left for safe-keeping. Although a limited and single-handed enquiry conducted by Mr AE Jeffreys between 1961 and 1964, with the support of the Commission and the Standing Conference of National and University Libraries, brought to light much information of value, it also made clear that only a sustained and more broadly based collective effort could be expected to succeed in doing so on the scale and with the precision of detail necessary for any published work. For this reason, the initial scope of the joint committee's enquiry was limited to a systematic search for surviving papers of 200 named scientists and technologists selected by the Royal Society on grounds of their individual achievements. As the enquiry gathered momentum it was found practicable to increase this figure progressively to its present total of 635. But the original criterion for inclusion has throughout remained unchanged, namely that the individuals concerned should all be recognised to have advanced significantly the state of knowledge in their respective fields through their scientific discoveries or inventions. In the case of engineers and technologists their innovations should have involved the application of entirely new principles.

The results here presented accordingly leave out of account the sometimes large accumulations of papers left by many antiquaries, speculative thinkers, explorers and others prominent in the world of scientific thought or the early affairs of the Royal Society, who could not themselves be regarded as original scientists. It is, however, hoped that further systematic exploration of these larger fields may become possible at a later date. Some indication of their size and complexity is given by Gavin DR Bridson, Valerie C Phillips and Anthony P Harvey, *Natural History Manuscript Resources in the British Isles* (Mansell,

1980), which also drew extensively for its information on the National Register of Archives, and by RM Macleod and JR Friday whose *Archives of British Men of Science* (Mansell, 1972, microfiche) records the results of an attempt to trace the descendants of some 3,000 nineteenth and twentieth-century scientists in a search for their surviving papers.

Following the normal practice of the series to which the Guide has now been assigned, it sets out to locate and describe all significant recorded groups of papers of the scientists with whom it deals. The terminal date of 1940 was selected with a view to excluding all living scientists. Scientists who died between 1940 and 1980 have been included if they had already carried out substantive work before the earlier date.

Administrative papers and correspondence among the records of government departments, official committees or learned institutions in whose work individual scientists may have participated has had normally to be ignored. Letters addressed to other individuals, and now included among the papers of the latter, have received separate notice only when they form substantial groups. Drawings have been treated on a similar basis.

Transcripts have been left unnoticed unless the originals from which they were made are lost, or the transcriber himself is the subject of the entry. Students' notes of the lectures of individual scientists have in principle also been ignored.

References to printed sources (indicated by the symbol ▷) or to unpublished lists of which the National Register of Archives holds copies (referred to by number, eg NRA 11420) have been given only when these amplify the information contained in the Guide itself. Holding institutions for which no location is indicated are situated in London.

The practicability of the venture has had, as always, to depend upon the cooperation and goodwill of innumerable owners and custodians of scientific papers throughout the world, as well as of individual scholars who have generously placed the results of their own researches at the Commission's or the joint committee's disposal. While their names are too numerous to recite, and can only be left to emerge from the text of the Guide itself, the immense value of their contribution requires the most grateful acknowledgement.

GRC DAVIS
Secretary
31 December 1981

Contents

Access to privately owned collections of papers

Privately owned collections of papers that have been deposited on loan by their owners in libraries, record offices and other public institutions are normally available for research without restriction. Special conditions, however, may sometimes apply to their use, particularly if they are to be cited in published works. All enquiries on this point should be addressed to the institution concerned.

Access to other privately owned collections remains entirely at the discretion of their owners, who have in some cases also asked that their identities should not be publicly disclosed. Enquiries about all collections whose location has been described simply as 'Private' should accordingly be addressed in the first instance to the Commission itself.

The manuscript papers of British scientists 1600-1940

---

[1] **ABEL, Sir Frederick Augustus,** 1st Bt (1827-1902)
FRS, chemist

Letters (34) to Sir GG Stokes 1864-89. *Cambridge Univ Libr* (in MS Add 7656)
Letters (8) to HE Armstrong 1884-95. *Imperial Coll, London* [NRA 11420]
Correspondence (6 items) with Sir HE Roscoe and the Chemical Society as organising secretary of the Imperial Institute 1882-91. *R Soc of Chemistry*

[2] **ABERNETHY, John** (1764-1831)
FRS, surgeon

Surgical lectures 1820-21; student notes (2 sets) of his lectures, 1802 and nd. *R Coll of Physicians, London*
Notes on suppuration in the cranium. *Edinburgh Univ Libr* (MS La II 157)
Papers (4) on human and whale anatomy read to the Royal Society 1793-98. *Royal Soc*
Student notes (14 sets) of his lectures 1806-21. *R Coll of Surgeons, London*

[3] **ACLAND, Sir Henry Wentworth,** 1st Bt (1815-1900)
FRS, physician

Correspondence and papers 1820-1900, incl diaries, a sketch book, accounts, and family, medical and personal correspondence. *Bodleian Libr, Oxford* (MSS Acland) [NRA 22893]
Miscellaneous correspondence and papers 1850-1900, incl a day book 1851-58 and reports concerning the Radcliffe Library 1856-62. *Radcliffe Science Libr, Oxford* [NRA 12664]
Correspondence (280ff) with WE Gladstone 1841-97. *Brit Libr* (Add MS 44091)
Correspondence (169ff) with Florence Nightingale 1867-97. *Ibid* (in Add MS 45786)
Letters (62) from him and his wife to Sir WC and Lady Trevelyan 1847-69. *Newcastle Univ Libr* [NRA 12238]
Family correspondence *c*1822-39. *Devon RO, Exeter* [NRA 14687]
Miscellaneous letters 1843-1900. *R Coll of Physicians, London*

See also Brodie BC the elder, Burdon-Sanderson, Gull, Huggins, Lister J, Owen, Paget, Rolleston, Simon J

[4] **ADAM, Neil Kensington** (1891-1973)
FRS, chemist

Correspondence and papers 1899-1970, incl 17 laboratory notebooks 1932-36, 37 other notebooks, and annotated copies of his own and other publications. *Southampton Univ Libr* [NRA 23476]
Correspondence with Viscount Cherwell rel to liquid hydrogen 1915. *Nuffield Coll, Oxford* [NRA 16447]

[5] **ADAMI, John George** (1862-1926)
FRS, pathologist

Papers and notes 1881-1926, incl the MS of his *Principles of pathology* (1909-10), a diary and papers as Colonel, Canadian Army Medical Corps and Medical Historical Recorder, Canadian Army in France 1915-18, and drawings and descriptions rel to influenza 1918-25. *Wellcome Hist Medical Libr* (MSS 846-53, 2417)
Miscellaneous letters 1919-26. *Liverpool Univ Archives*
Miscellaneous correspondence. *Medical Research Council*

[6] **ADAMS, John Couch** (1819-1892)
FRS, astronomer

Papers (1 vol) on Uranus 1841-46, diaries 1855-89, and letters addressed to him. *St John's Coll, Cambridge* [NRA 9502]
Correspondence, mainly with his family, and papers 1828-90, incl mathematical and astronomical notes, graphs and plans, poems, diplomas, awards and testimonials. *Cornwall RO, Truro* [NRA 24521]

Report to the Royal Astronomical Society 1849 and 109 letters to its officers 1845-90; letters (31) to the Revd R Sheepshanks 1846-53. *R Astronomical Soc*
Letters (27) to Sir JW Lubbock 1848-59, and miscellaneous letters and papers. *Royal Soc*
Letters (36) to Sir GG Stokes 1849-*c*1890. *Cambridge Univ Libr* (in MS Add 7656)
Correspondence (20 items) with Sir GB Airy 1849-73. *R Greenwich Observatory, Herstmonceux* (in MSS 941-53)

[7] **ADRIAN, Edgar Douglas,**
1st Baron Adrian of Cambridge, OM
(1889-1977)
FRS, physiologist

Fellowship dissertation 1913, speeches (4 vols) as Master of Trinity College 1951-65, and letters to Sir GI Taylor 1911-74. *Trinity Coll, Cambridge* [NRA 22889]
Miscellaneous correspondence and papers, mainly rel to research topics and administration. *Medical Research Council*
Correspondence with Viscount Cherwell 1919-55. *Nuffield Coll, Oxford* [NRA 16447]

[8] **AIRY, Sir George Biddell,** KCB
(1801-1892)
FRS, astronomer and mathematician

Correspondence and papers as astronomer royal, mainly rel to the administration and work of the Royal Observatory and associated government business. *R Greenwich Observatory, Herstmonceux* (MSS 597-1333) [NRA 22822]
Papers (21) submitted to the Royal Astronomical Society 1842-75; 325 letters to its officers 1832-88 and 214 to its secretary the Revd R Sheepshanks 1836-59. *R Astronomical Soc*
Letters (370) to Sir GG Stokes 1848-82, with 10 letters to the Revd TR Robinson 1830-71. *Cambridge Univ Libr* (in MS Add 7656)
Correspondence (327 items) with Sir JFW Herschel 1824-75; letters (113) to Sir JW Lubbock 1829-60; miscellaneous letters (220). *Royal Soc*
Correspondence (187 items) with JD and G Forbes 1832-80. *St Andrews Univ Libr* [NRA 13132]
Correspondence (118 items) with Sir JFW Herschel 1836-70. *Texas Univ, Austin*
Letters (39) to J Tyndall and others. *Royal Inst*
Letters (24) to Sir D Gill 1875-79. *R Geographical Soc*
Letters (22) to S Grimaldi 1837-60. *Brit Libr* (in Add MS 34189)
Letters (20) to C Babbage 1826-55. *Ibid* (in Add MSS 37183-86, 37194, 37196)
Correspondence with the Earl of Ellenborough 1845-46. *Public Record Office* (PRO 30/12/4/21) [NRA 21870]
Letters (15) to M Faraday 1835-61. *Inst of Electrical Engineers* (Blaikley Colln)
Miscellaneous letters and papers. *Trinity Coll, Cambridge*
See also Adams, Baily, Beaufort, Carrington, De La Rue, De Morgan, Faraday, Forbes JD, Henslow, Herschel JFW, Johnson MJ, Peacock, Sabine, Sedgwick, Sheepshanks, Smyth, Stephenson R, Stewart, Stokes, Stratford, Whewell

[9] **AITCHISON, James Edward Tierney**
(1835-1898)
FRS, naturalist

'Afghan plants' (1 vol) 1875; notes (8 packets) on the uses, native names etc of plants of the Punjab, W Himalayas, Sind, Afghanistan, Baluchistan and NE Persia for a projected work 'Florae Indiae desertae'. *R Botanic Gardens, Kew*

[10] **ALDER, Joshua** (1792-1867)
Zoologist

Notebook on mollusca *c*1835-64. *Brit Mus (Nat Hist) Zoology Dept*
Correspondence 1826-67. *Ibid, General Libr*
Letters to GB Sowerby and sketches. *Nat Mus of Wales Zoology Dept, Cardiff*
Drawings of mollusca and other marine animals. *Hancock Mus, Newcastle upon Tyne*

[11] **ALEXANDER, Wilfred Backhouse**
(1885-1965)
Ornithologist

Papers, incl a bibliography (2 vols) of British birds, and correspondence rel to his collaboration with DL Lack 1941-43. *Edward Grey Inst, Zoology Dept, Oxford*

[12] **ALLAN, Thomas** (1777-1833)
FRS, mineralogist

His catalogue of his collection of minerals (3 vols). *Brit Mus (Nat Hist) Mineralogy Dept*

[13] **ALLMAN, George James** (1812-1898)
FRS, botanist and zoologist

Letters (18) to TH Huxley 1851-81. *Imperial Coll, London*

[14] **ALSTON, Charles** (1683-1760)
Physician and botanist

Papers, incl a treatise (3 vols) and lectures on materia medica and botany 1733-36, notes of experiments 1738-50, and 46 letters addressed to him 1720-60. *Edinburgh Univ Libr* (MSS Gen 659-97; MS Dc 8.12; MS La III 375)

Lectures on botany 1724, catalogue of the Edinburgh botanic garden 1743, index of plants wintering in the greenhouse 1743, 'Hippocratis medicamenta'. *R Botanic Garden, Edinburgh*
Treatise (4 vols) on materia medica 1740. *R Coll of Surgeons, London*
Lectures (3 vols) on materia medica 1720-54. *Medical Soc of London* (MS 98)

See also Fothergill

[15] **ANDERSON, Sir Hugh Kerr** (1865-1928)
FRS, physiologist

Scientific and family papers and correspondence. *Cambridge Univ Libr* (MS Add 7649)
Correspondence and papers. *Medical Research Council*

[16] **ANDERSON, John** (1726-1796)
FRS, natural philosopher

Papers and correspondence, incl his lecture notes, rain gauge records (1 vol) 1792, working copy of his *Institutes of physics* (1786), MSS of other published works, commonplace books (3 vols) and tour of the Hebrides. *Andersonian Libr, Strathclyde Univ, Glasgow*

▷ J Muir, *John Anderson, pioneer of technical education and the college he founded*, 1950, pp80-82

[17] **ANDERSON, William** (1750-1778)
Surgeon and naturalist

Notes on birds, animals and plants observed during the second and third voyages of Captain Cook 1772-78. *Brit Mus (Nat Hist) Botany Dept* (Banksian MS 81)
Notes on birds observed on Captain Cook's second voyage 1772-75. *Ibid, Zoology Dept*

[18] **ANDREWES, Sir Frederick William** (1859-1932)
FRS, pathologist and bacteriologist

Papers, incl notes on pathology 1897-1906, experiments on iodine method for catgut ligatures 1916, serological research at St Bartholomew's Hospital 1918. *Wellcome Hist Medical Libr* (MSS 952-53)
Miscellaneous correspondence mainly rel to research topics. *Medical Research Council*

[19] **ANDREWS, Charles William** (1866-1924)
FRS, palaeontologist

Notebooks (4) kept on Christmas Island 1897-98; report to Sir J Murray on phosphate of lime deposits there; 23 letters to CD Sherborn 1897-1908. *British Mus (Nat Hist) General Libr*

[20] **ANDREWS, Thomas** (1813-1885)
FRS, chemist

Papers and correspondence. *Queen's Univ, Belfast*
Notes on experiments (8 vols) 1836-77; correspondence (10 items) with Sir JFW Herschel. *Royal Soc*
Correspondence (2 vols) 1842-73. *Science Mus Libr*
Letters (35) to Sir GG Stokes 1860-83. *Cambridge Univ Libr* (in MS Add 7656)
Correspondence (18 items) with Lord Kelvin 1846-77. *Cambridge Univ Libr* (in MS Add 7342)
Correspondence (14 items) with Lord O'Hagan 1867-81. *Public Record Office of Northern Ireland, Belfast* (D 2777/7/A/1-14) [NRA 18813]
Correspondence with Sir BC Brodie the younger *c*1870-74. *History of Science Mus, Oxford* (MS Mus 66)

[21] **APPLETON, Sir Edward Victor,** GBE (1892-1965)
FRS, physicist

Correspondence and papers, incl diaries, notebooks, speeches and broadcasts. *Edinburgh Univ Libr* [NRA 24825]
Letters to Sir OW Richardson 1924-36. *Texas Univ, Austin* [NRA 22337]
Correspondence with Viscount Cherwell 1940-55. *Nuffield Coll, Oxford* [NRA 16447]
Miscellaneous correspondence. *Medical Research Council*

[22] **ARDERON, William** (1703-1767)
FRS, naturalist

Correspondence and papers, incl notes, specimens and drawings (6 vols) on natural history and geology 1742-64, and correspondence (*c*1,200 items in 4 vols) with H Baker 1744-67. *Victoria & Albert Mus* (Forster Colln)
Correspondence and papers (351ff) mainly on natural history, geology, physics, shorthand and Norwich trades, dialect and architecture 1742-67. *Brit Libr* (Add MS 27966)
Notes and articles on arts and trades 1728-50, lecture notes (3 vols) taken by him 1728, and an unpublished novel. *Norfolk RO*
Rainfall measurements at Norwich 1753. *Brit Libr* (in Add MS 4439)
Drawings (1 vol) of fossils found in Norfolk 1749-52, with other drawings 1741-49. *Castle Mus, Norwich*
Papers (22) submitted to the Royal Society. *Royal Soc*

▷PES Whalley, 'William Arderon, FRS, of Norwich', *J Soc Bibliog Nat Hist*, vi, pt 1, 1971, pp30-49

[23] **ARKELL, William Joscelyn** (1904-1958)
FRS, geologist

Correspondence and papers 1933-58, incl MSS of his published work and notes, correspondence and annotated maps on the Jurassic system in Great Britain. *University Mus, Oxford* (Geological Collns) [NRA 24520]
Correspondence with JW Tutcher. *Bristol Mus and Art Gallery*

[24] **ARKWRIGHT, Sir Richard**
(1732-1792)
Engineer

Business papers, incl correspondence and payrolls, together with those of his son Richard 1782-1815. *Columbia Univ Libraries, New York* (Seligman Papers)
Arkwright Cotton Mill wages books (4 vols) 1786-1811. *Chesterfield Public Libr*
Miscellaneous papers, mainly rel to his Derbyshire estates *c*1782-92. *Brit Libr* (in Add MSS 6668-97)

[25] **ARMSTRONG, Henry Edward**
(1848-1937)
FRS, chemist and educationalist

Personal and family correspondence and papers, incl notes on the teaching of science, and diary of his visit to the USA with the Mosely Commission 1903. *Imperial Coll, London* [NRA 11420]
Miscellaneous letters (40). *Royal Soc*

See also Abel, Crookes, Dewar, Lodge, Roscoe

[26] **ARMSTRONG, William George,**
Baron Armstrong of Cragside (1810-1900)
FRS, inventor

Business records of Sir WG Armstrong & Co and successor companies. *Tyne and Wear Archives Dept, Newcastle upon Tyne*
Business records of Sir WG Armstrong & Co and successor companies. *Private* [NRA 23464]
Letters (15) to TH Huxley 1874-92. *Imperial Coll, London*
Correspondence (11 items) with Sir GG Stokes 1869-87. *Cambridge Univ Libr* (in MS Add 7656)

[27] **ARNOTT, George Arnott Walker**
(1799-1868)
Botanist

Correspondence, incl letters (21 vols) from other botanists, letters (6 vols) to TG Rylands 1860-68, and letters to C Johnson, RK Greville and R Taylor. *Brit Mus (Nat Hist) Botany Dept*
Letters (406) to Sir WJ Hooker 1825-64. *R Botanic Gardens, Kew*
Letters (49) to Margaret Gatty 1853-63. *Sheffield Central Libr* (MD 2132) [NRA 15568]

AVEBURY, Baron, see Lubbock, J

[28] **BABBAGE, Charles** (1792-1871)
FRS, mathematician

Correspondence (20 vols) 1806-71 and papers (4 vols), incl essays on the philosophy of analysis, astronomy lectures, and notes on astronomy, mechanical drawing, lighthouses, geology, cyphers and mathematical recreations. *Brit Libr* (Add MSS 37182-37205)
'Notations' (8 vols), 15 notebooks, and *c*250 machine drawings. *Science Mus Libr*
Papers 1808-*c*1866, incl experimental notebooks and notes on mathematics, astronomy, physics, electricity etc. *Scientific Periodicals Libr, Cambridge* [NRA 9501]
Papers (6) submitted to the Royal Society 1815-26; correspondence (402 items) with Sir JFW Herschel 1812-66; 25 letters to Sir JW Lubbock 1829-60. *Royal Soc*
Papers (5) on the calculus of functions and signs in mathematical reasoning 1815-21; 32 letters to Mary and W Somerville 1828-37. *Bodleian Libr, Oxford* (Somerville Colln, MS B-1, 16) [NRA 9423]
Papers, incl notes and drawings of engines, a paper on the science of numbers reduced to mechanism, proofs of his *Passages from the life of a philosopher* (1864). *History of Science Mus, Oxford* (MSS Buxton 3-14)
Copy of his *Passages from the life of a philosopher* (1864) interleaved with letters addressed to him 1804-47. *McGill Univ Libr, Montreal*
Miscellaneous papers and correspondence, incl notes on mathematical tables and 6 letters to Sir GG Stokes 1855-69. *Cambridge Univ Libr* (in MS Add 7656)
Letters (72) to Sir J Franklin. *Scott Polar Research Inst, Cambridge*
Correspondence (23 items) with Sir JFW Herschel 1814-63. *Texas Univ, Austin* [NRA 13677]
Letters, incl 10 to officers of the Royal Astronomical Society 1820-70, and 7 to A De Morgan *c*1835-51. *R Astronomical Soc*
Letters (15) to A Quételet 1826-62. *Bibl Royale, Brussels*
Correspondence (14 items) with JD Forbes 1831-55. *St Andrews Univ Libr* [NRA 13132]
Letters (13) to the Duke of Somerset 1830-49.

Buckinghamshire RO, Aylesbury (Bulstrode Papers)

See also Airy, Baily, Brewster, Brodie BC the elder, Brunel IK, Brunel MI, De Morgan, Donkin, Fairbairn, Faraday, Fitton, Forbes JD, Gilbert, Harcourt WVV, Harris, Henry, Herschel JFW, Ivory, Lubbock JW, Lyell, Murchison, Peacock, Playfair, Robinson TR, Sedgwick, Somerville, Sylvester, Talbot, Wheatstone, Whewell, Whitworth

[29] **BABINGTON, Charles Cardale** (1808-1895)
FRS, botanist

Correspondence and papers incl the MS (3 vols) of his *Flora of Cambridgeshire* 1845-60 and annotated copies of his own and other published works. *Botany School, Cambridge*
Insect collection catalogue 1876. *Mus of Zoology, Cambridge*
Letters (25) to Sir WJ Hooker 1834-51. *R Botanic Gardens, Kew*

▷AM Babington, *Memorials, journal, and botanical correspondence of Charles Cardale Babington*, 1897

[30] **BAILLIE, Matthew** (1761-1823)
FRS, morbid anatomist

Miscellaneous papers incl case notes 1790, papers on paraplegia in adults 1822, and transcripts of his autobiographical memoranda. *R Coll of Physicians, London*
European travel journal 1788; student notes on his anatomy lectures 1794-99. *R Coll of Surgeons, London* [NRA 9521]

See also Clift

[31] **BAILY, Francis** (1774-1844)
FRS, astronomer

Correspondence and papers *c*1798-1844, incl notebooks, observations and computations rel to star catalogues and the mean density of the earth, general catalogue of the principal stars, tract on fluxions, notebook on a Hadley's sextant, papers submitted to the Royal Astronomical Society and abstracts prepared for the Society's *Monthly Notices. R Astronomical Soc*
Pendulum observations, correspondence (261 items) with Sir JFW Herschel 1820-44, 48 letters to Sir E Sabine and 44 to Sir JW Lubbock 1830-40. *Royal Soc*
Tables etc *c*1812-44; letters addressed to him *c*1822-43. *London Univ Libr* [NRA 9193]
Letters (44) to C Babbage 1820-35. *Brit Libr* (in Add MSS 37182-86, 37188-89)
Letters (26) to SP Rigaud 1830-38. *Bodleian Libr, Oxford* (MS Rigaud 60)
Letters (20) to Sir GB Airy 1835-43. *R Greenwich Observatory, Herstmonceux* (in MS 938)
Correspondence (10 items) with JD Forbes 1837-*c*1844. *St Andrews Univ Libr* [NRA 13132]

See also Brisbane, Robinson TR

[32] **BAINBRIDGE, John** (1582-1643)
Physician and astronomer

Astronomical and mathematical papers. *Bodleian Libr, Oxford* (MS Smith 92; MS Add A 380)
Astronomical, mathematical and chronological collections and calculations, catalogue of mathematical instruments and correspondence with scientists. *Trinity Coll, Dublin* (MSS 382-86, 794)

[33] **BAKER, John Gilbert** (1834-1920)
FRS, botanist

Catalogues of Madagascan plants (4 vols), papers on ferns (4 boxes) and correspondence. *R Botanic Gardens, Kew*
Notes (1 vol) on plant records of the Lake District and Yorkshire 1865-82; notes (1 vol) on the life and character of J Fothergill. *Brit Mus (Nat Hist) Botany Dept*
'Flora of the Lake District' (2 vols). *Cumbria County Libr, Carlisle*

[34] **BAKER, William** (1787-1853)
Naturalist

Correspondence and papers (352ff) rel to natural history 1811-48, incl notes and drawings. *Brit Libr* (Add MS 35173)
His natural history lecture notes. *Bridgwater Public Libr* (Maurice Page MSS) [NRA 9186]
Letters (11) from WH Dixon 1851-52. *Brit Libr* (in Add MS 35058)

See also Buckland

[35] **BALL, John** (1818-1889)
FRS, man of science

Correspondence and papers, incl a catalogue of Moroccan plants (3 vols) 1890, 'Distribution of plants on the south side of the Alps' (2 vols) 1896, and 28 letters to Sir WJ Hooker 1842-55. *R Botanic Gardens, Kew*
Correspondence (23 items) with the Marquess of Ripon 1871-87. *Brit Libr* (in Add MS 43545)

[36] **BALL, Sir Robert Stawell** (1840-1913)
FRS, astronomer and mathematician

Correspondence rel to Dunsink Observatory, Dublin, 1885-1900. *History of Science Mus, Oxford* (MS Mus 54) [NRA 9532]

Letters (96) to officers of the Royal Astronomical Society 1872-1900. *R Astronomical Soc*
Letters (17) to Sir GG Stokes 1875-*c*1898. *Cambridge Univ Libr* (in MS Add 7656)
Letters 1893-95. *Trinity Coll, Dublin*

[37] **BANKS, Sir Joseph,** Bt (1743-1820) FRS, naturalist

[The dispersal of Banks's own papers and the present location of the more important groups are described by JC Beaglehole, *Endeavour journal of Sir Joseph Banks*, I, 1962, pp127-39. The British Museum (Natural History) holds large quantities of transcripts, microfilms and photocopies of his correspondence and journals assembled by Dawson Turner (no.586 below), Warren R Dawson and others]

Correspondence and papers, mainly foreign correspondence 1765-1820. *Brit Libr* (Add MSS 8094-8100, 8967-68, 33977-82, 52281)
Correspondence (1 vol) and papers, incl a book of weights of his friends and acquaintances 1778-1814. *Brit Mus (Nat Hist) Botany Dept*
Correspondence and papers, incl memoranda and drafts of his letters 1764-1810, letters addressed to him (4 vols) 1766-80, and transcripts of his journals 1766-71. *R Botanic Gardens, Kew*
Correspondence and papers, incl botanical notes and drawings (1 vol), journal of his voyage to Iceland 1772, transcripts of his journals 1763-71, and business and estate papers 1755-1820. *Kent Archives Office, Maidstone* (Knatchbull Papers) [NRA 1301]
Transcript of his Hebridean journal 1772, and papers rel to his estates in Lincolnshire. *Ibid* (Stanhope Papers)
Correspondence and papers (9,148 items) mainly rel to sheep farming, fen drainage, coinage, botany, entomology, Iceland, St Helena, Africa, India and the East and West Indies 1780-1820. *Sutro Libr, San Francisco*
Correspondence and papers (25 vols with others unbound) mainly rel to the exploration of Australia and the Pacific, Captain Bligh, botany, horticulture, and the Lincolnshire shrievalty 1766-1820; his HMS *Endeavour* journal (2 vols) 1768-71. *Mitchell Libr, Sydney* [NRA 19108]
Journal (2 vols) of his voyage to Newfoundland and Labrador 1766. *R Geographical Soc of Australia, Adelaide*
His HMS *Endeavour* journal (3 vols) 1768-71. *Derby Central Libr* (MSS 2722-24) [NRA 20300]
Journal of his voyage to Iceland 1772, transcript of his HMS *Endeavour* journal 1768-71, and notes on the natural history of Newfoundland 1766. *Blacker-Wood Libr, McGill Univ, Montreal*
Journal of his tour in Holland 1773. *Wellcome Hist Medical Libr* (MS 1049)
Journal of his tour in Holland 1773, diplomas 1785-1813, and correspondence 1768-1819. *Dixson Libr, Sydney* (MS 101, MSS Q 158-61)
His notes on Cook's voyages. *Public Libr, Auckland* (Grey MSS 47-75)
Description of Tahiti 1769 and other papers 1767-1806. *Alexander Turnbull Libr, Wellington*
Vocabularies of Pacific and other languages collected by him *c*1770-80. *School of Oriental and African Studies, London*
Poems (70 items) and correspondence mainly rel to Iceland (76 items) 1772-1819. *Univ of Wisconsin Libr, Madison*
Correspondence and papers (*c*3,500 items) mainly rel to agriculture and his estates in Lincolnshire and Middlesex 1763-1819. *Sterling Memorial and Beinecke Libraries, Yale Univ, New Haven*
Correspondence and papers mainly rel to agriculture, estate and family affairs. *Lincolnshire Archives Office* [NRA 4675, 5348, 6329, 9511]
Correspondence and papers mainly rel to his estates in Lincolnshire 1790-1812. *Brit Libr* (Add MS 43837)
Papers rel to his estates, agriculture, vermin, decoys, fen drainage etc, with report on American corn pests 1788. *Spalding Gentlemen's Soc*
Papers and correspondence mainly rel to the militia in Lincolnshire. *Lincolnshire County Libr* [NRA 5342]
Journals of his tours to Bristol, Wales, the Midlands etc 1767-68. *Cambridge Univ Libr* (MS Add 6294)
Correspondence about Derbyshire lead mining 1793-1800. *Imperial Coll, London* (Robert Annan Mining Colln)
Papers rel to the British Museum, with *c*230 miscellaneous items of correspondence down to 1810. *Fitzwilliam Mus, Cambridge* (SG Perceval Colln)
Papers rel to the Royal Society, and family correspondence and papers. *Royal Soc*
Miscellaneous correspondence (*c*231 items) down to 1820. *Nat Libr of Australia, Canberra*
Miscellaneous correspondence (78 items) 1780-1820. *American Philosophical Soc, Philadelphia*
Correspondence (59 items) with Sir W Hamilton 1777-1803. *Brit Libr* (in Add MS 34048; in Egerton MS 2641)
Letters (67) to J Lloyd *c*1778-1814. *Nat Libr of Wales* (MS 12415)
Letters (52) to Sir JE Smith 1786-1820, with gardening hints for the embassy to China 1792. *Linnean Soc*

Letters (47) to the Duke and Duchess of Somerset 1813-19. *Buckinghamshire RO, Aylesbury* [NRA 11704]

▷ *The Banks letters: a calendar of the manuscript correspondence preserved in the British Museum, the British Museum (Natural History) and other collections in Great Britain*, ed WR Dawson, 1958, with supplements published in *Bull Brit Mus (Nat Hist) historical ser*, iii, pt 2, 1962, pp41-70, and iii, pt 3, 1965, pp71-93, and PM Teiger, 'Supplementary letters of Sir Joseph Banks, third series', *J Soc Bibliog Nat Hist*, vii, pt 3, 1975, pp249-57; G Meynell, 'Banks papers in the Kent Archives Office, including notebooks by Joseph Banks and Francis Bauer', *Archives of Nat Hist*, x, pt i, 1981, pp77-88

See also Blagden, Brown R, Daubeny, Dryander, Herschel W, Home, Hope J, Knight, Maskelyne, Pennant, Roxburgh, Thompson B, Wallich N, Young A

[38] **BARCROFT, Sir Joseph** (1872-1947)
FRS, physiologist

Papers 1915-35. *Physiology Dept, Cambridge*
Correspondence with and concerning him (*c*60 items) 1928-47. *Medical Research Council* (file 1563)
Correspondence (23 items) with AV Hill 1919-47. *Churchill Coll, Cambridge*
Correspondence with Viscount Cherwell and related papers concerning experiments on dogs 1940-41. *Nuffield Coll, Oxford* [NRA 16447]

[39] **BARKLA, Charles Glover** (1877-1944)
FRS, physicist

Papers and correspondence, incl drafts of articles and letters rel to his receipt of the Nobel Prize. *Private* [NRA 22850]
Letters (*c*17) to Sir OW Richardson *c*1929-34. *Texas Univ, Austin*
Letters (10) to Lord Rutherford 1909-21. *Cambridge Univ Libr* (in MS Add 7653)

[40] **BARLOW, Peter** (1776-1862)
FRS, mathematician

Report and correspondence with T Telford concerning the London Bridge project 1823. *Inst of Civil Engineers* [NRA 14021]
Correspondence (12 items) with Sir JFW Herschel; letters (12) to Sir JW Lubbock 1830-36. *Royal Soc* [NRA 8803]
Letters to the Revd TJ Hussey. *University Coll London* (MSS Graves 23A, 23B)

[41] **BARRETT, Sir William Fletcher** (1844-1925)
FRS, physicist

Working papers and letters addressed to him. *Royal Soc*
Correspondence (102 items) with Sir OJ Lodge 1889-1925. *Incorporated Soc for Psychical Research* [NRA 11857]
Letters (14) to Sir GG Stokes 1867-99. *Cambridge Univ Libr* (in MS Add 7656)

[42] **BARROW, Revd Isaac** (1630-1677)
FRS, mathematician

Papers on Apollonius, Archimedes and others. *Royal Soc*
Letters (17) to J Collins 1663-70. *Private*

[43] **BATES, Henry Walter** (1825-1892)
FRS, naturalist

Pocket book 1848-59. *Brit Libr* (Add MS 42138)
Illustrated notebooks (2) on the insects of the Amazon basin 1851-59. *Brit Mus (Nat Hist) Entomology Dept*
Letters (39) to CR Darwin 1861-79. *Cambridge Univ Libr*
Correspondence (15 items) with Sir C Lyell 1863-73. *Edinburgh Univ Libr* (in MS Lyell 1)

See also Darwin CR

[44] **BATESON, William** (1861-1926)
FRS, biologist

Correspondence and papers, incl lecture notes 1897-1904, notebooks, working papers for *Materials for the study of variation* (1894) and its unpublished sequel, and correspondence with C Dobell, Sir F Galton, CC Hurst, Sir JS Huxley, TH Huxley, Sir ER Lankester, K Pearson, Sir EP Poulton, RC Punnett and the Revd A Sedgwick. *Cambridge Univ Libr*
Record books of experimental work, with RC Punnett, on lathyrus from 1903 and poultry from 1908. *Genetics Dept, Cambridge*
Papers as director of the John Innes Horticultural Institution, Merton, Surrey (1910-26). *John Innes Inst, Norwich*
Correspondence (*c*30 items) with Sir F Galton *c*1888-1909. *University Coll London* [NRA 19968]
Letters (13) to Sir J Larmor. *St John's Coll, Cambridge*
Letters (10) to K Pearson 1896-1912. *University Coll London*

▷ W Coleman, 'Bateson papers', *The Mendel Newsletter: archival resources for the history of genetics and allied sciences, issued by the Library of the American Philosophical Society*, ii, Nov 1968, pp1-3 and AG Cock,

'The William Bateson papers', *ibid*, xiv, June 1977, pp1-4

[45] **BATHER, Francis Arthur** (1863-1934)
FRS, geologist

Notebooks and drawings incl sketches illustrating papers on cephalod and crinoid morphology 1892; notes on the Henry Johnson collection 1913. *Brit Mus (Nat Hist) Palaeontology Dept*

[46] **BAXENDELL, Joseph** (1815-1887)
FRS, meteorologist and astronomer

Astronomical observations etc 1836-87, incl his Southport observatory journal 1877-87, variable star charts and notes, 3 papers submitted to the Royal Astronomical Society and 31 letters to its officers 1857-87. *R Astronomical Soc*

[47] **BAXTER, William** (1787-1871)
Botanist

Report on the Fielding Herbarium and the Sherard Room 1865-70, plant lists and catalogues, botanical notes. *Bodleian Libr, Oxford* (MSS Sherard 22-25, 377-79, 381-88) [NRA 6305]

'Flora aquatica Oxoniensis' 1825. *Botany School, Cambridge*

Notes on customs and superstitions rel to trees and plants. *Bodleian Libr, Oxford* (MS Eng misc c 90)

Annotated copy of J Ray's *Synopsis methodica stirpium Britannicarum* (1690). *Botany Dept, Oxford*

Annotated copy of J Ray's *Synopsis*, 3rd edn (1724). *Taylor Inst, Oxford*

[48] **BEAUFORT, Sir Francis,** KCB (1774-1857)
FRS, hydrographer

Correspondence (*c*1,050 items) and papers 1787-1857, incl journals (5) 1791-1822, notes for his *Karamania* (1817), and drawings of Mediterranean ports. *Huntington Libr, San Marino* (Brit Libr microfilm RP 276)

Report on Monteith's map of Armenia 1831; paper on proposed expedition to Wager Bay, Australia 1836; letters to Sir G Back *c*1848-57. *R Geographical Soc*

Papers rel to his testimonial, and correspondence (26 items) with Sir GB Airy 1850-57. *R Greenwich Observatory, Herstmonceux* (in MSS 942-47)

Correspondence (115 items) with Sir JFW Herschel 1828-52; letters (38) to Sir JW Lubbock 1830-45. *Royal Soc*

Letters (*c*100) to the Society for the Diffusion of Useful Knowledge 1827-45. *University Coll London*

Letters (76) to Sir WJ Hooker 1829-57. *R Botanic Gardens, Kew*

Miscellaneous correspondence (20 items) 1837-57. *Scott Polar Research Inst, Cambridge* [NRA 6527]

[49] **BELL, Alexander Graham** (1847-1922)
Inventor

Papers and correspondence (*c*130,000 items), incl his laboratory notebooks, family papers and photographs, and drawings of his telephone. *Libr of Congress, Washington*

Papers and correspondence rel to his inventions 1890-1922. *Alexander Graham Bell National Historic Park, Baddeck, Nova Scotia*

'Beinn Bhreagh Recorder', a journal of experiments with his inventions and animal husbandry kept by him and his associates 1909-22. *Private*

Correspondence (30 items) with T Borthwick 1878-85. *Nat Libr of Scotland* (Acc 4005)

[50] **BELL, Sir Charles** (1774-1842)
FRS, surgeon

Watercolour drawings (17) made in 1836 of battle wounds received at Waterloo; letters (5) to him on this subject 1815. *Royal Army Medical Coll, London* (95, 630) [NRA 5981]

Illustrated manuscript on the anatomy of the brain 1823. *Wellcome Hist Medical Libr* (MS 1121)

Catalogue of his museum of anatomy and surgery, with related papers 1825-26. *R Coll of Surgeons, Edinburgh* [NRA 20261]

Letters (75) 1825-37, incl 19 to Lord Brougham 1825-37 and 17 to the Society for the Diffusion of Useful Knowledge 1827-33. *University Coll London* (College Correspondence, Brougham Papers, SDUK Papers)

Journal of an Italian tour 1840. *London Univ Libr* [NRA 9193]

[51] **BELT, Thomas** (1832-1878)
Geologist

Letters (180) addressed to him 1864-78. *Brit Mus (Nat Hist) General Libr*

MS on Northumberland plants 1848-51. *Hancock Mus, Newcastle upon Tyne*

Notes and diagrams of Thames Valley gravels. *Brit Mus (Nat Hist) Palaeontology Dept*

[52] **BENNETT, George** (1804-1893)
Naturalist and physician

Papers, incl an account of his specimens of comparative anatomy and natural history presented to the museum of the Royal College of Surgeons of England, and notes on the habits of the sperm whale;

correspondence (30 items) with Sir R Owen 1833-40. *R Coll of Surgeons, London*

Journal (3 vols) of his world travels 1877-79. *Mitchell Libr, Sydney*

Letters to Sir R Owen 1844-88. *Brit Mus (Nat Hist) General Libr*

Letters (13) to Sir WJ Hooker 1853-64. *R Botanic Gardens, Kew*

[53] **BENTHAM, George** (1800-1884)
FRS, botanist

Correspondence and papers, incl 'Leguminosae' (9 vols), 'Phytogeography', diary (20 vols) 1807-83, autobiography (2 vols), memoirs of botanists, notes, catalogues, indexes, letters addressed to him (10 vols), correspondence (1,220 items) with Sir WJ Hooker 1823-65; correspondence with Sir FJH von Müller *c*1858-84. *R Botanic Gardens, Kew*

Diary 1853, letters (*c*200) to his family 1804-53, and miscellaneous papers 1820-83. *Linnean Soc* [NRA 9516]

Letters (250) to A Gray 1839-83. *Arnold Arboretum and Gray Herbarium Libr, Harvard Univ, Cambridge, Mass*

Letters (116) to Sir FJH von Müller 1858-77. *R Botanic Gardens, South Yarra, Australia*

See also Seemann

[54] **BERKELEY, Revd Miles Joseph** (1803-1889)
FRS, mycologist

Correspondence (11 vols), papers and 530 water colour drawings of fungi. *Brit Mus (Nat Hist) Botany Dept*

Index to his mycological herbarium; drawings (4 vols); letters (143) to Sir WJ Hooker 1832-65. *R Botanic Gardens, Kew*

Papers (15) read to the Linnean Society 1839-74, miscellaneous notes and drawings. *Linnean Soc*

[55] **BERKELEY, Randal Mowbray Thomas Rawdon,** 8th Earl of Berkeley (1865-1942)
FRS, physicist

Laboratory notebooks (46) *c*1903-1928, and letters addressed to him *c*1895-1928. *Private*

Letter book containing copies of correspondence about apparatus for his laboratory 1911-15; letters to the Revd FJ Jervis-Smith. *History of Science Mus, Oxford* (MSS Mus 53, 62) [NRA 9532]

Correspondence with Viscount Cherwell 1928-35. *Nuffield Coll, Oxford* [NRA 16447]

[56] **BERNAL, John Desmond** (1901-1971)
FRS, crystallographer

Scientific and political papers and correspondence (117 boxes), incl published and unpublished writings on science and on the social relations of science, material for his books, diaries and notebooks. *Cambridge Univ Libr*

Papers rel to his international peace activities. *London School of Economics*

Correspondence with I Fankuchan *c*1936-64. *American Inst of Physics, New York*

Correspondence with WT Astbury 1941-54. *Brotherton Libr, Leeds Univ*

[57] **BERNARD, Revd Edward** (1638-1696)
FRS, astronomer

Papers and correspondence, incl 'Chronicon omnis aevi', MS of *Catalogi librorum manuscriptorum* (1697), essays on mathematics and poetry, oriental and western vocabularies and grammars, commentaries on the Bible, devotional papers, and notes on mathematics, astronomy, chronology, medicine, logic, etymology, ancient alphabets, ancient history, Jewish antiquities and theology. *Bodleian Libr, Oxford*

See also Flamsteed

[58] **BESSEMER, Sir Henry** (1813-1898)
FRS, engineer and inventor

Papers and correspondence concerning the Dowlais Iron Co 1856-97. *Glamorgan Archive Service, Cardiff* (D/DG/C2) [NRA 9506]

Correspondence rel to the Bessemer medal, Royal School of Mines 1881. *Imperial Coll, London*

[59] **BLACK, Joseph** (1728-1799)
Chemist

Correspondence (*c*490 items) 1748-99, and autobiographical notes. *Edinburgh Univ Libr* (MSS Gen 873-75) [NRA 13028]

Correspondence (79 items) with J Watt 1768-99. *Private* [NRA 22549]

Correspondence (*c*70 items) with J Watt *c*1768-99. *Birmingham Reference Libr*

Correspondence (22 items) 1782-92. *Private* (photocopies in Public Record Office of Northern Ireland, ref T 1703)

Letters (12) to A Black 1784-90. *Private*

See also Hope TC, Richardson BW

[60] **BLACKETT, Patrick Maynard Stuart,** Baron Blackett, OM (1897-1974)
FRS, physicist

Papers and correspondence 1920-74, incl notebooks, reports, papers rel to

World War II, and MSS of his published work, lectures and broadcasts. *Royal Soc* [NRA 22627]
Correspondence (30 items) with N Bohr 1930-61. *Niels Bohr Inst, Copenhagen*
Correspondence with S Gill 1964-69. *Science Mus Libr* [NRA 21399]
Correspondence with Sir RE Peierls 1930-54. *Bodleian Libr, Oxford* [NRA 20805]
Correspondence with EC Stoner 1937-62. *Brotherton Libr, Leeds Univ* [NRA 17735]
Letters to Lord Jackson 1957-69. *Imperial Coll, London*

See also Cockcroft, Simon FE, Tizard

[61] **BLAGDEN, Sir Charles** (1748-1820)
FRS, physician

Correspondence and papers (1,033 items) *c*1770-1820, incl notes on meteors, earthquakes etc, travel notes, literary memoranda, diaries 1776-88, letter book 1783-87; letters (98) to Lord and Lady Palmerston 1788-1804. *Beinecke Libr, Yale Univ, New Haven* (Osborn Colln: Blagden Papers and MSS c 114, f c 15-16) [NRA 18661]
Papers 1767-*c*1780, incl clinical notes, case book, notes on lectures attended, commonplace book, lists of materia medica, and memoirs (2) addressed to the President of the Royal Society. *Wellcome Hist Medical Libr* (MSS 1234-52)
Correspondence (1,215 items) and papers, incl a diary 1771-1820. *Royal Soc*
Miscellaneous correspondence and papers 1775-1819, incl family letters (19), daily accounts, vouchers and receipts. *Gloucestershire RO* (D 1086) [NRA 7013]
Letters (122) to Sir J Banks 1786-1802. *Brit Libr* (Add MS 33272)
Letters (75) to Sir J Banks 1773-1805. *Fitzwilliam Mus, Cambridge* (SG Perceval Colln)
Correspondence (41 items) with Sir W Herschel 1781-1814. *R Astronomical Soc*
Letters (18) to Sir J Banks 1784-85. *R Botanic Gardens, Kew*

[62] **BLAIR, Patrick** (*c*1666-1728)
FRS, physician and botanist

Correspondence and papers, incl anatomical and chirurgical observations, descriptions of an elephant, seal and porpoise, petition to the Royal College of Physicians, 63 letters to Sir H Sloane 1705-28 and 30 to J Petiver 1708-15. *Brit Libr* (in Sloane MSS 3321-22, 3812, 4020, 4025, 4040-49, 4058, 4065-66)
'Medicinal, chirurgicall and anatomicall observations'; 'The anatomy of the seal and porpess'; paper on the structure of plants read to the Royal Society 1720; 8 letters from him 1712-27. *Royal Soc*
Preface to an unfinished treatise; copies of his correspondence 1721-25. *Bodleian Libr, Oxford* (in 4° Rawl 323)
Copies of his correspondence 1725-27; catalogue of the botanical discoveries made by him. *Brit Mus (Nat Hist) Botany Dept* (Banksian MS 35)

[63] **BLAND, Revd Miles** (1786-1867)
FRS, mathematician

Miscellaneous papers (15 items) 1802-39. *Liverpool Univ Libr*
Correspondence (7 items) with Sir JFW Herschel. *Royal Soc* [NRA 8803]

[64] **BLOXAM, Revd Andrew** (1801-1878)
Naturalist

Natural history notes made on the voyage of HMS *Blonde* to South America 1824-25. *Brit Mus (Nat Hist) General Libr*
Notes on the cellular cryptogams of Leicestershire, and correspondence. *Ibid, Botany Dept*
MS of 'Flora Leicesterensis' written with WH Coleman 1852 and published after their deaths as Coleman's *Flora of Leicestershire* (1886). *Leicestershire RO*
Letters (11) to Sir WJ Hooker 1839-56. *R Botanic Gardens, Kew*

[65] **BOBART, Jacob** (1641-1719)
Botanist

Botanical notebooks (8) incl analyses of R Morison and J Bobart's *Plantarum historia Oxoniensis*, pts ii and iii (1680-99) and botanical lists. *Bodleian Libr, Oxford* (MSS Lat misc c 11, d 25-27, e 28-31)
Papers and correspondence incl catalogues of plants in Oxford and plants of medical use, interleaved and annotated copy of *Catalogus horti botanici Oxoniensis* by P Stephens and W Browne (1658), miscellaneous plant lists and notes. *Ibid* (MSS Sherard 29-42) [NRA 6305]
Working copy of C Bauhin's *Pinax theatri botanici* (1625). *Brit Mus (Nat Hist) Botany Dept*
Letters (26) to J Petiver 1700-16, and 15 to Sir H Sloane 1685-1716. *Brit Libr* (in Sloane MSS 3321-22, 4036-38, 4040-41, 4043-44, 4064)
Letters (14) to W Sherard. *Royal Soc*

See also Dillenius

[66] **BONE, William Arthur** (1871-1938)
FRS, chemical technologist

Analysis books (4 vols) 1890-1912, administrative correspondence, especially rel to the Chemical Technology Department of Imperial College 1911-30, and biographical papers. *Imperial Coll, London*

[67] **BOOLE, George** (1815-1864)
FRS, mathematician

Scientific papers (5 boxes). *Royal Soc*
Papers on symbolic logic. *University Coll Dublin* (MS Q.5.44)
Correspondence (*c*150 items). *Private*
Correspondence (83 items) with A De Morgan 1842-64. *University Coll London* (MS Add 97) [NRA 14258]
Letters (35) to Lord Kelvin 1845-55. *Cambridge Univ Libr* (in MS Add 7342)
Letters (15) from Lord Kelvin 1845-48. *Glasgow Univ Libr*

[68] **BOOTT, Francis** (1792-1863)
Physician and botanist

Botanical journal (4 vols) 1814-*c*1832; drawings, proofs and correspondence rel to his *Illustrations of the genus carex* (1858-67); notes on herbaria 1817-19; letters (230) to Sir WJ Hooker 1818-63. *R Botanic Gardens, Kew*
Botanical journal (3 vols) 1816-19. *Hunt Institute for Botanical Documentation, Carnegie-Mellon Univ, Pittsburgh*
Letters (200) to A Gray 1839-63. *Arnold Arboretum and Gray Herbarium Libr, Harvard Univ, Cambridge, Mass*
Letters (26) to Sir JE Smith 1816-*c*1821. *Linnean Soc*

[69] **BORRER, William** (1781-1862)
FRS, botanist

Correspondence and papers, incl botanical notes 1851-55, letters (145) to Sir WJ Hooker, and 21 volumes of transcripts and photocopies of his correspondence. *R Botanic Gardens, Kew*
Botanical notes 1837-46; letters (8) to Sir JE Smith 1806-26. *Linnean Soc*
Correspondence. *Botany School, Cambridge*

See also Spruce, Turner D

[70] **BOWER, Frederick Orpen** (1855-1948)
FRS, botanist

Correspondence and papers *c*1885-1938, incl lecture notes, class registers, biological drawings and diagrams (1 vol), papers rel to the Royal Botanic Institution of Glasgow and botanic gardens. *Glasgow Univ Archives* [NRA 18906]

See also Darwin F, Seward

[71] **BOYCOTT, Arthur Edwin** (1877-1938)
FRS, pathologist and naturalist

Notebooks, files and correspondence on mollusca. *Brit Mus (Nat Hist) Zoology Dept*
Miscellaneous correspondence rel to research topics and administration. *Medical Research Council*

[72] **BOYLE, Hon Robert** (1627-1691)
FRS, natural philosopher and chemist

Papers arranged by subject under science (15 vols), theology (12 vols), philosophy (4 vols), physiology (3 vols) and miscellaneous (12 vols), notebooks (16), correspondence (7 vols) and miscellaneous letters (175). *Royal Soc*
Letters (13) to his father and others 1640-83, with accounts and other papers rel to his Irish estates, and geometrical problems by him. *Private* [NRA 20594/15]
Statements (4) as witness of cures performed by V Greatrakes 1666. *Brit Libr* (in Add MS 4293)
Correspondence (19 items) with the Commissioners of the United Colonies of New England 1662-84. *Guildhall Libr, London* (MS 7936)
Transcripts of letters (20) to the Revd N Marsh respecting the publication of the Irish Bible 1682-84 (originals destroyed). *Marsh's Libr, Dublin* (MS Z 4.4.8)
Transcripts of letters (18) to the Revd N Marsh respecting the publication of the Irish Bible (originals destroyed). *John Rylands Univ Libr, Manchester* (Eng MS 502)
Miscellaneous letters and papers. *Nat Libr of Ireland* (Orrery MSS 32-36, 7163-76; Ormonde MS 87)

▷REW Maddison, 'A tentative index of the correspondence of the Honourable Robert Boyle, FRS', *Notes and Records*, xiii, 1958, pp128-201

See also Oldenburg

[73] **BRADLEY, Revd James** (1693-1762)
FRS, astronomer

Correspondence (2 vols) and papers 1717-62, incl lectures on optics, astronomy, hydrostatics etc, astronomical observations, computations and tables, and business papers. *Bodleian Libr, Oxford* (MSS Bradley 1-22, 25-38, 40-45; MSS Rigaud 34, 48, 50, 56; MS Eng misc e 15)
Lectures (20) on physics; algebra and geometry notebook; astronomical treatises (3 vols). *University Coll London* (MSS Graves 3-5c)
Observations, computations, etc. *R Greenwich Observatory, Herstmonceux* (MSS 90-135) [NRA 22822]
Observations on Halley's comet 1757. *Brit Libr* (in Add MS 4439)

See also Hornsby

[74] **BRADLEY, Richard** (1688-1732)
FRS, botanist

Letters (13) to J Petiver 1714-15 and 9 to Sir H Sloane 1714-28. *Brit Libr* (in Sloane MSS 1968, 3322, 4045-49, 4058, 4065)

[75] **BRAGG, Sir William Henry,** OM, KBE (1862-1942)
FRS, physicist

Correspondence and papers (*c*10,000 items), incl diaries, working notebooks, lectures, speeches, broadcasts, and copy of unpublished autobiography. *Royal Inst*
Notes (1 vol) on his research with Sir WL Bragg on X-rays 1913. *Brotherton Libr, Leeds Univ* (MS 81)
Correspondence (65 items) with Lord Rutherford 1904-33. *Cambridge Univ Libr* (in MS Add 7653)
Letters (19) to Sir OW Richardson 1924-36. *Texas Univ, Austin* [NRA 22337]
Miscellaneous letters (15) 1907-31. *Royal Soc*

[76] **BRAGG, Sir William Lawrence** (1890-1971)
FRS, physicist

Correspondence and papers (*c*30,000 items), incl working notebooks, lectures, speeches and broadcasts. *Royal Inst*
Papers 1914-18. *R Artillery Inst* (MD/364)
Correspondence with Viscount Cherwell 1919-56. *Nuffield Coll, Oxford* [NRA 16447]
Correspondence with EC Stoner 1932-64. *Brotherton Libr, Leeds Univ* [NRA 17735]
Miscellaneous correspondence. *Medical Research Council*

See also Bragg WH

[77] **BRANDE, William Thomas** (1788-1866)
FRS, chemist

Laboratory notebooks (2) 1821-59, and miscellaneous letters (18). *Royal Inst*
Letters (10) addressed to him 1839-57. *R Coll of Physicians, London*
Correspondence (28 items) with Sir JFW Herschel 1850-59. *Texas Univ, Austin*
Letters (12) to Sir JW Lubbock 1836-42. *Royal Soc*

See also Faraday

[78] **BREWSTER, Sir David** (1781-1868)
FRS, natural philosopher

Diplomas, awards and related correspondence 1800-67, with patent and specification for his kaleidoscope 1817. *Private* [NRA 16036]
MS of his *Report on certain optical and other instruments* (1856), and 88 letters to WHF Talbot 1839-65. *Science Mus Libr*
MS of a paper on crystalline reflections. *Manchester Central Libr*
Correspondence and papers rel to university administration. *St Andrews Univ Libr*
Correspondence, incl *c*200 letters to Lord Brougham 1827-66 and 25 to officers of the Society for the Diffusion of Useful Knowledge 1827-32. *University Coll London* (Brougham Papers, SDUK Papers and MS Graves 23A)
Correspondence (218 items) with JD Forbes 1828-67. *St Andrews Univ Libr*
Correspondence with J Lee 1809-27, Messrs Blackwood 1817-63, J Skene 1819-31, WH Lizars 1832-59, and others 1808-68. *Nat Libr of Scotland*
Correspondence (107 items) with C Babbage 1818-67. *Brit Libr* (in Add MSS 37182-37200)
Letters (65) to J Veitch 1797-1832. *Private* [NRA 11619]
Correspondence (53 items) with Sir JFW Herschel 1817-67. *Royal Soc*
Letters (50) to M Napier 1821-44. *Brit Libr* (in Add MSS 34612-24)
Letters (*c*25) from him and his wife to Sir WC and Lady Trevelyan 1840-68. *Newcastle Univ Libr* [NRA 12238]
Letters (22) to SP Rigaud 1820-38. *Bodleian Libr, Oxford* (in MS Rigaud 60)
Letters (18) to Sir GG Stokes 1848-67. *Cambridge Univ Libr* (in MS Add 7656)
Letters (12) to Lord Playfair 1851-68. *Imperial Coll, London* [NRA 11556]

[79] **BRISBANE, Sir Thomas Makdougall,** 1st Bt (1773-1860)
FRS, astronomer

Correspondence and papers (3 vols), incl papers on family and personal matters 1794-1860 and military affairs 1792-1857, observations made at Makerston Observatory, and correspondence 1802-59. *Mitchell Libr, Sydney* (MS 1191)
Correspondence 1812-37 and papers 1815-58, incl reports, addresses received 1825, and autobiographical fragments and notes. *Nat Libr of Australia, Canberra* (MS 4036, NK 6787)
Correspondence and papers (64 items) 1813-15, incl reports and memoranda. *William L Clements Libr, Univ of Michigan, Ann Arbor*
Correspondence and papers, incl transit observations 1823, papers (3) submitted to the Royal Astronomical Society 1822-25, 13 letters to its officers 1820-58 and 15 to F Baily *c*1830. *R Astronomical Soc*
Family and estate correspondence, accounts and other papers. *Mitchell Libr, Glasgow* [NRA 24533]

Correspondence and papers 1819-26. *Private* [NRA 10544]
Miscellaneous correspondence and papers *c*1807-49. *Private* [NRA 11854]
Correspondence (26 items) with JD Forbes 1831-59. *St Andrews Univ Libr* [NRA 13132]

See also Ure

[80] **BRODIE, Sir Benjamin Collins,** the elder, 1st Bt (1783-1862)
FRS, surgeon

Papers 1802-54, incl case notes (19 vols) 1805-54, hospital notes (4 vols) 1813-16, physiological experiments and observations (3 vols) 1810-26, surgical cases and commentaries (2 vols) 1805-07, 'Psychologia' 1851, commonplace book, essays, and student notes of his lectures. *St George's Hospital Medical School Libr, London* [NRA 9767]
Papers, incl autobiography, receipt book for fees 1816-23, notes (679ff) for his physiology lectures 1820-23 and student notes 1810. *R Coll of Surgeons, London*
Letters and papers, incl prescriptions 1829-41, testimonial given by him 1854, and notes accompanying specimen for museum 1824. *R Coll of Physicians, London*
Letters (89ff) to Sir HW Acland 1843-62 and testimonial for Acland 1847. *Bodleian Libr, Oxford* [NRA 22893]
Letters (37) to Sir W Sharpey 1854-62. *University Coll London*
Correspondence about the Bill for the regulation of medical practice 1841-52. *Radcliffe Science Libr, Oxford* (MS ZB 14)
Letters (15) to C Babbage 1834-61. *Brit Libr* (in Add MSS 37188-98)
Letters (14) to Sir JW Lubbock 1833-60; correspondence (12 items) with Sir JFW Herschel 1859-61. *Royal Soc*
Miscellaneous letters (12). *Royal Inst*

[81] **BRODIE, Sir Benjamin Collins,** the younger, 2nd Bt (1817-1880)
FRS, chemist

Correspondence 1870-74 and papers, incl laboratory notebooks, MS and corrected proofs of his *Ideal chemistry. A lecture* (1880), other corrected proofs, and student notes of his lectures 1861-62. *History of Science Mus, Oxford* (MSS Mus 66-79, 103) [NRA 9532]
Correspondence (38 items), mainly on the calculus of chemical operations, 1853-78, and papers, incl his report of a conversation about Sir H Davy 1839, a few literary essays and poems. *Leicester Univ Libr* [NRA 9991]
Letters (81) to Sir GG Stokes 1862-80. *Cambridge Univ Libr* (in MS Add 7656)
Letters to Sir J Alderson and testimonials received 1867-68. *R Coll of Physicians, London*

See also Andrews T

[82] **BROMFIELD, William Arnold** (1801-1851)
Botanist

Papers and correspondence, incl American and West Indian journals, notes on plants and insects found in Europe 1826-31, notes on Hampshire and Egyptian plants 1849-51, annotated set of his tracts on Hampshire botany, MS (10 vols) of his *Flora Vectensis* (1856); letters (45) to Sir WJ Hooker 1836-50. *R Botanic Gardens, Kew*

[83] **BROWN, John** (1780-1859)
Geologist

Notes on English post-Pliocene deposits and their fossils *c*1841-1858, and some letters. *Brit Mus (Nat Hist) Palaeontology Dept*

[84] **BROWN, Robert** (1773-1858)
FRS, botanist

Papers, incl diaries (2) 1800-05, notes on botanical specimens collected during the voyage of HMS *Investigator* 1801-05, slip catalogue (34 boxes) describing plants in his herbarium and that of Sir J Banks, flora of New Holland and of Madeira, and indexes to his general correspondence (3 vols). *Brit Mus (Nat Hist) Botany Dept*
Notes on zoological specimens collected during the voyage of HMS *Investigator* 1801-05. *Ibid, Zoology Dept*
Notes on mineral specimens collected during the voyage of HMS *Investigator* 1801-05. *Ibid, Mineralogy Dept*
Papers and correspondence 1800-58; notes and correspondence on the composition of water *c*1839-1855. *Brit Libr* (Add MSS 32439-41, 33227)
Papers submitted to the Linnean Society, list of New Holland bird specimens presented by him to the Society 1818, and 14 letters to Sir JE Smith 1806-25. *Linnean Soc*
Letters (46) to Sir WJ Hooker 1820-53. *R Botanic Gardens, Kew*

▷PI Edwards, 'Robert Brown (1773-1858) and the natural history of Matthew Flinders' voyage in HMS *Investigator*, 1801-1805', *J Soc Bibliog Nat Hist*, vii, pt 4, 1976, pp 385-407

See also Curtis J, Griffith

[85] **BROWN, Samuel** (1817-1856)
Chemist

Essays, poems, family papers and correspondence 1833-56. *Nat Libr of Scotland* (MSS 1889-90)

[86] **BRUCE, Sir David,** KCB (1855-1931)
FRS, pathologist

Reports on tsetse fly disease in Zululand 1896, 1903; South African diary 1900. *Royal Army Medical Coll, London* (MSS 564, 569)
MS of 'Malta fever' 1890. *Wellcome Hist Medical Libr* (MS 1375)
Letters rel to the Croonian lectures 1911-15. *R Coll of Physicians, London*

[87] **BRUNEL, Isambard Kingdom** (1806-1859)
FRS, civil engineer

Diaries, account books, sketch books, and 49 letters. *Bristol Univ Libr*
Journal (2 vols) 1830-34, with a few additions down to 1840. *Private*
Memorandum and letter books (2 vols) rel to various engineering projects 1842-51. *Huntington Libr, San Marino*
Letter books. *Newcastle Univ Libr*
Reports, letter books, prospectuses etc on building the Great Western Railway *c*1833-59; letters on building the South Devon and Cornwall Railways. *Public Record Office* (British Transport Historical Records)
Great Western Railway correspondence 1846-49 and plans (2) 1853. *Devon RO, Exeter* (Kennaway Papers) [NRA 9463]
Cornwall Railway correspondence 1843-58. *Cornwall RO, Truro* (Treffry of Place Papers) [NRA 5005]
Letters (26) addressed to him 1834-51. *Bodleian Libr, Oxford* (MS Autogr c 6)
Letters (37) to C Babbage 1827-57. *Brit Libr* (in Add MSS 37184-94, 37196-97, 37200)
Letters (16) to officers of the Society for the Diffusion of Useful Knowledge 1830-43. *University Coll London* [NRA 22241]

[88] **BRUNEL, Sir Marc Isambard** (1769-1849)
FRS, civil engineer

Letter book rel to his block-making machinery 1806-11. *Nat Maritime Mus, Greenwich* (LBK 54)
Papers (11) mainly rel to the Thames tunnel 1818-42. *Science Mus Libr*
Letters (17) to C Babbage 1823-42, with notes on 'Brunel's machine' 1831. *Brit Libr* (Add MSS 37183-92 *passim*)
Letters (13) to Earl Spencer 1816-34. *Private*
Letters (2) with scale drawings of timber carriages 1811. *Nat Libr of Scotland* (Acc 4202)

[89] **BRUNTON, Sir Thomas Lauder,** 1st Bt (1844-1916)
FRS, physician

Notebooks, correspondence and papers (10 vols) *c*1864-83; lectures (2 vols) on therapeutics 1892; notes for his lectures on chloroform, eye affections, alcohol and drugs *c*1895. *Wellcome Hist Medical Libr* (MSS 1384-86, 5966-75)
Letters to Macmillan & Co *c*1875-1916. *Brit Libr* (Add MS 55247)

[90] **BUCHANAN** (afterwards **HAMILTON**), **Francis** (1762-1829)
FRS, surgeon and naturalist

Papers, incl extracts made by him from his Burma journal 1795, observations on Nepal 1802-03, notes for and MS of his *Account of the fishes found in the river Ganges* (1822), and papers rel to his survey of Bengal 1807-14, comprising reports (21 vols), statistics (14 vols), vocabularies (5 vols), drawings, maps and inscriptions (5 vols). *India Office Libr and Records* (MSS Eur C 12-14, D 70-98, 541, E 68-73, G 10-25; MS Kurukh D 1; NHD 2-3)
Papers, incl his Burma journal (2 vols) and list of plants observed on his journey to Ava 1795, botanical notes and papers (1 vol) 1798-1801, and notes and commentaries on botanical publications (5 vols). *Edinburgh Univ Libr* (MSS Dc 1. 11-14, 71-74, MS Dc 6.97)
Correspondence and reports on leprosy 1797-98, notes on Indian flora 1795-97, list of stones brought from Indian 1815, genealogical papers, and correspondence with naturalists 1802-29. *Scottish Record Office* (GD 161/19) [NRA 8142]
Papers rel to the flora and fauna of India and Nepal, incl field notes, lists of seeds, animal and plant descriptions (314), an incomplete flora of Nepal, drawings, and 32 letters to Sir JE Smith 1783-1825. *Linnean Soc*
List of plants observed on his journey to Ava 1795, with drawings (53), catalogue of plant specimens collected in Chittagong and Tripura 1798, and letters (1 vol) to W Roxburgh 1795-1812. *Brit Mus (Nat Hist) Botany Dept*
Descriptions of fishes found in the river Ganges and of a journey through Chittagong and Tripura 1798. *Brit Libr* (Add MSS 9882, 19296)

See also Roxburgh

[91] **BUCHANAN, John Young** (1844-1925)
FRS, chemist

MS 'On the physical and chemical work of an Antarctic expedition' 1899. *R Geographical Soc*
Personal correspondence (60 items) 1870-1908, incl 20 letters to his parents from on board HMS *Challenger* 1872-75. *Private* [NRA 16920, 17389]

[92] **BUCKLAND, Revd William** (1784-1856)
FRS, geologist

Scientific, personal and family correspondence and papers. *Devon RO, Exeter* (Gordon and Buckland Colln) [NRA 11695]
The MS of parts of his *Geology and mineralogy considered with reference to natural theology* (1836), 12 packets of lecture notes 1820-50 and *c*50 letters addressed to him. *University Mus, Oxford* (Geological Collns)
Annotated copy of A de Caumont's *Carte géologique du dept de la Manche, partie du nord* (1827), annotated drawings of fossil cephalopods, ammonites and belemnites 1840, student notes of his lectures, and miscellaneous letters 1823-49. *Inst of Geological Sciences*
Geological notes. *Geology Dept, Oxford*
Geological sketches and notes; correspondence (50 items) with W Baker 1841-52; 11 letters to T Webster 1814-32. *Fitzwilliam Mus, Cambridge* (SG Perceval Colln)
Correspondence (*c*300 items) 1818-54. *Royal Soc*
Correspondence (*c*260 items) with Sir R Peel 1823-48. *Brit Libr* (Add MSS 40355-40600 *passim*)
Letters (55) from Sir R Peel 1836-49. *Beinecke Libr, Yale Univ, New Haven* (Osborn Colln, MS d 61) [NRA 18661]
Letters (29) from JB Pentland 1820-22. *Nottingham Univ Libr* [NRA 24710]
Letters (30) from Sir RI Murchison. *Geological Soc of London*
Correspondence (*c*50 items) with Lord Playfair, Sir R Peel and others. *Imperial Coll, London* [NRA 11556]
Correspondence (33 items) with Lord and Lady Grenville 1818-31. *British Libr* (in Add MS 58995)
Letters (64ff) to the Marquess of Northampton 1831-49. *Private* [NRA 21088]
Letters (60ff) 1833-49. *Bodleian Libr, Oxford* (in MS Eng lett d 5)
Letters (20) to Sir HT De la Beche. *Nat Mus of Wales Geology Dept, Cardiff*
Letters (17) to Sir WJ Hooker 1832. *R Botanic Gardens, Kew*
Letters (17) to Sir WC and Sir J Trevelyan. *Newcastle Univ Libr* [NRA 12238]
Letters (15) to Sir R Owen. *R Coll of Surgeons, London*

See also Mantell, Murchison, Owen, Woodward SP

[93] **BUCKTON, George Bowdler** (1818-1905)
FRS, entomologist

Drawings for his *Monograph of the British aphides* (1876-83), with some unclassified drawings *c*1876; annotated watercolour drawings (354) of tettigidae. *Brit Mus (Nat Hist) Entomology Dept*
Drawings (3 vols) for his *Monograph of the membracidae* (1903). *Hope Libr, University Mus, Oxford*

[94] **BUDDLE, Revd Adam** (*c*1660-1715)
Botanist

Botanical papers, catalogues and memoranda, and correspondence (14 items) with J Petiver *c*1700. *Brit Libr* (in Sloane MSS 2201, 2305-06, 2970-80, 4063, 4066)

[95] **BULLER, Arthur Henry Reginald** (1874-1944)
FRS, botanist

Botanical notebooks (3 vols), proofs of illustrations for and MS of an unpublished eighth volume of his *Researches on fungi*, and MSS of 'A history of mycology' and 'The fungus lore of the Greeks and Romans'. *R Botanic Gardens, Kew*

[96] **BURCHELL, William John** (1781-1863)
Naturalist

Catalogues of plants (14 vols) of Portugal and the Canary Islands, Africa and Brazil, flora of St Helena and related drawings (3 vols), botanical memoranda and plant lists incl his 'Hortus Fulhamensis' (4 vols), miscellaneous notebooks (7 boxes); letters (12) to Sir WJ Hooker 1819-47. *R Botanic Gardens, Kew*
Entomological papers. *Hope Libr, University Mus, Oxford*
Letters (23) to W Swainson 1819-39. *Linnean Soc* [NRA 22111]

[97] **BURDON-SANDERSON, Sir John Scott,** 1st Bt (1828-1905)
FRS, physiologist

Correspondence 1853-1905 and papers, incl diaries (44 vols) 1861-1905, notes and drafts of lectures etc, and family papers. *University Coll London* (MS Add 179; Galton Papers) [NRA 14259, 19968]
Clinical notebooks (6 vols) 1864-70, revised copies of his teaching works, vivisection licences 1876-1905, and correspondence (38 items) with CR Darwin. *Woodward Biomedical Libr, Univ of British Columbia, Vancouver* (Sinclair Colln) [NRA 14062]
Report on EC Baber's paper on the thyroid of the dog 1876 and some letters, incl 14 to CR Darwin and 4 to Sir GG Stokes. *Cambridge Univ Libr* (in MS Add 7656; Darwin Papers) [NRA 11458]
Letters to Sir HW Acland 1881-96. *Bodleian Libr, Oxford* [NRA 22893]

See also Lister JJ

[98] **BUSK, George** (1807-1886)
FRS, man of science

Notebooks, drawings and MSS rel to his works on polyzoa 1852-86. *Brit Mus (Nat Hist) Zoology Dept*
Papers, incl notes, lectures, diary and notes of tours in France and Italy. *R Coll of Surgeons, London*
Lithographic plates (38) of human crania. *Brit Mus (Nat Hist) Palaeontology Dept*
Letters (51) to Margaret Gatty 1851-67. *Sheffield Central Libr* (MD 2131) [NRA 15568]

[99] **CANTLIE, Sir James,** KBE (1851-1926)
Surgeon

Lectures, diaries, notebooks, case books, cash book, reports, MSS of published and unpublished work and other papers 1876-1911. *Wellcome Hist Medical Libr* (MSS 1456-99, 4780)

[100] **CANTON, John** (1718-1772)
FRS, natural philosopher

Correspondence and papers (3 vols); notes rel to his pneumatic experiments and Canton committee minutes. *Royal Soc*
Papers (28 items) rel to his experiments on the compressibility of water for the Royal Society's Canton committee. *Scottish Record Office* (GD 150/2647)

See also Priestley

[101] **CARLIER, Edmond William Wace** (1861-1940)
Physiologist

Lecture notes, notebooks, notes, MSS and corrected proofs of his published work, and correspondence (714 items) 1875-1939. *Birmingham Univ Libr* [NRA 14359]

See also Lodge

[102] **CARPENTER, Geoffrey Douglas Hale** (1882-1953)
Entomologist

Entomological notes and correspondence. *Hope Libr, University Mus, Oxford*
Correspondence with WE Waller. *Cumberland House Mus, Portsmouth*

[103] **CARPENTER, Philip Herbert** (1852-1891)
FRS, palaeontologist and zoologist

Correspondence 1884-85 rel to his *Catalogue of the blastoidea in the British Museum* (1886). *Brit Mus (Nat Hist) Palaeontology Dept*
Notes on TH Huxley's lectures on natural history 1869-71. *Zoology Dept Libr, Oxford*

[104] **CARPENTER, William Benjamin** (1813-1885)
FRS, naturalist

Materials (3 vols) for an unpublished monograph on the eozoon. *Brit Mus (Nat Hist) Palaeontology Dept*
Drawings (54) and papers rel to his published work on the microscopic structure of shells. *Ibid, Zoology Dept*
Geology lectures 1848. *Brit Libr* (Add MS 31196)
Correspondence (353ff) with the Earl and Countess of Lovelace 1843-45 and (88ff) with W Greig 1845. *Bodleian Libr, Oxford* (MSS Dep Lovelace Byron 65, 169) [NRA 21980]
Letters (23) to TH Huxley 1851-83. *Imperial Coll, London*
Letters (17) to Sir JFW Herschel 1838-71. *Royal Soc*
Letters (13) to Sir GG Stokes 1855-79. *Cambridge Univ Libr* (in MS Add 7656)
Letters (11) to Sir C Lyell 1850-74. *Edinburgh Univ Libr* (in MS Lyell 1)

[105] **CARRINGTON, Richard Christopher** (1826-1875)
FRS, astronomer

Papers, incl collections of problems in mathematics, optics and astronomy (6 vols) 1842-48, transit-circle ledgers (5 vols) 1853-56, Red Hill catalogue of zone corrections (3 vols), diagrams (24) of atmospheric pressure and temperature, Durham 1850-51, and unpublished translations of works by JL Lagrange and GB Libri 1848. *R Observatory, Edinburgh* [NRA 9504]
Sunspot observations (14 vols) 1853-73, papers submitted to the Royal Astronomical Society and letters, incl 43 to officers of the Society 1851-74. *R Astronomical Soc*
Papers (2) projecting Bessel's star-zones from 1825 to 1875 and Argelander's star-zones from 1850 to 1875; correspondence (21 items) with Sir JFW Herschel 1857-64. *Royal Soc*
Correspondence (17 items) with Sir GB Airy 1854-67. *R Greenwich Observatory, Herstmonceux* (in MSS 945-51)

[106] **CARTWRIGHT, Revd Edmund** (1743-1823)
FRS, inventor

Patents, specifications and plans of wool-combing machines, legal correspondence and other papers 1788-1813. *Gloucestershire RO* (D 1245/CF 1-9) [NRA 5851]

[107] **CATTON, Revd Thomas** (1760-1838)
FRS, astronomer

Astronomical notes 1823. *Cambridge Univ Libr* (MS Add 6229)
Letters (2) to officers of the Royal Astronomical Society 1821, 1833. *R Astronomical Soc*

[108] **CAVALLO, Tiberius** (1749-1809)
FRS, natural philosopher

Correspondence and papers, incl correspondence (21 items) with the Revd T Rackett, Count Zenobio, PS Andreadi, J Lind and others 1786-1809, notes on a portable barometer, cement, and wax candles, business agreements rel to the publication of his *Elements of natural and experimental philosophy* (1803), inventory and sale catalogue of his library, mathematical instruments and other effects, and executorship papers 1810-28. *Dorset RO* (Solly Papers) [NRA 8506]
Letters (2 vols) to J Lind 1782-1809. *Brit Libr* (Add MSS 22897-98)
Letters (21) to M van Marum 1788-1802. *Netherlands Soc of Sciences, Haarlem*
Miscellaneous letters and papers (20 items). *Royal Soc*

[109] **CAVENDISH, Henry** (1731-1810)
FRS, natural philosopher

Scientific papers incl experimental notebooks and other notes rel to electricity, chemistry, physics, meteorology, optics, mathematics, dynamics and mechanics. *Private* [NRA 20594/14]
Letters and papers (32 items) rel to experiments on air. *Royal Soc*
Draft and fair copy (90 pp) of an unpublished paper on his discoveries regarding heat c1780. *Public Archives of Canada, Ottawa*

[110] **CAYLEY, Arthur** (1821-1895)
FRS, mathematician

Papers (2) submitted to the Royal Astronomical Society 1862 and nd; letters (144) to its officers 1864-91. *R Astronomical Soc*
Papers on Hansen's lunar theory and on postulation 1855. *Reading Univ Libr* (MS 139) [NRA 14186]
Correspondence (700 items) with JJ Sylvester 1847-93. *St John's Coll, Cambridge*
Correspondence (92 items) with Sir GG Stokes 1849-94. *Cambridge Univ Libr* (in MS Add 7656)
Letters (48) to TA Hirst c1860-90. *University Coll London* (London Mathematical Society Papers)
Letters (45) to Lord Kelvin 1845-94. *Cambridge Univ Libr* (in MS Add 7342)

[111] **CAYLEY, Sir George,** 6th Bt (1773-1857)
Pioneer of aerial navigation

Correspondence and papers incl notebooks and drawings. *R Aeronautical Soc* [NRA 1055]
▷ *Aeronautical and miscellaneous note-book, ca. 1799-1826, of Sir George Cayley. With . . . a list of the Cayley papers,* ed JE Hodgson, Newcomen Society, extra pubn 3, 1933

[112] **CHADWICK, Sir James** (1891-1974)
FRS, physicist

Correspondence and papers 1914-73, incl laboratory and other notebooks, engagement diaries, and papers rel to Lord Rutherford. *Churchill Coll, Cambridge*
Correspondence and papers rel to his work on atomic energy 1940-74, incl reports, engagement diaries and committee papers. *Atomic Energy Research Establishment, Harwell*
Correspondence (28 items) with Lord Rutherford 1914-37. *Cambridge Univ Libr* (in MS Add 7653)
Letters (16) from N Bohr 1930-61. *Niels Bohr Inst, Copenhagen*
Correspondence with Viscount Cherwell 1940-56. *Nuffield Coll, Oxford* [NRA 16447]

See also Cockcroft, Rutherford

[113] **CHAIN, Sir Ernst Boris** (1906-1979)
FRS, chemist

Scientific and personal papers and correspondence. *Private*

CHERWELL, Viscount, see Lindemann

[114] **CHEYNE, Sir William Watson,** 1st Bt (1852-1932)
Bacteriologist and surgeon

Hunterian Oration and other papers on war wounds 1915; fragment of war diary 1917. *Wellcome Hist Medical Libr* (MS 1591)

[115] **CHRISTIE, Alexander Turnbull** (1801-1832)
Surgeon

Lecture notes, travel journals and other papers. *Nat Libr of Scotland* (Acc 3442)

[116] **CHURCH, Arthur Harry** (1865-1937)
FRS, botanist

Papers, incl the MS and drawings for his *Types of floral mechanism* (1908), and some letters. *Brit Mus (Nat Hist) Botany Dept*

[117] **CHURCH, Sir Arthur Herbert,** KCVO (1834-1915)
FRS, chemist

Papers, incl lecture drafts, unpublished notes and drawings, and unpublished MSS on British algae. *Botany School, Cambridge*
Revised proof copy of his *Chemistry of paints and painting* (1890); drafts (2) of a paper on cleaning a fresco; letters (87) from Lord Leighton 1880-95. *R Academy of Arts* (LEI/44; MIS/CHU/1-3)

[118] **CLARK, Sir Wilfrid Edward Le Gros** (1895-1971)
FRS, anatomist

Correspondence and papers, incl diaries and journals (37) 1937-71 and drafts of articles. *Bodleian Libr, Oxford* [NRA 18598]
Correspondence (*c*150 items) with KP Oakley 1948-69. *Brit Mus (Nat Hist) General Libr*
Miscellaneous correspondence mainly rel to research topics and administration. *Medical Research Council*

[119] **CLARK, Revd William** (1788-1869)
FRS, anatomist

Letters from him *c*1807-21. *Scottish Record Office* (in GD 1/378) [NRA 9960]

[120] **CLARKE, Charles Baron** (1832-1906)
FRS, botanist

'Acanthaceae of South East Asia' (2 vols); 'Cyperaceae' (37 vols); notebooks and papers (9 vols). *R Botanic Gardens, Kew*
'Determinations of Wallich's herbarium – gesneraceae, bignoniaceae and pedalineae' 1881. *Linnean Soc*

[121] **CLARKE, Revd William Branwhite** (1798-1878)
FRS, geologist

Papers and correspondence 1827-78, incl geological field books, notebooks, sketch books, diaries, and meteorological records. *Mitchell Libr, Sydney* (ML MSS 141, 368, 454, 490; ML DOC 1395)
Notebooks 1852-53, papers and sketches. *Sydney Univ Libr*
Paper submitted to the Royal Society 1876; correspondence with Sir JFW Herschel and others. *Royal Soc*
Correspondence 1848-78. *Dixson Libr, Sydney*
Letters to A Sedgwick *c*1840-54. *Cambridge Univ Libr* (in MS Add 7652)
Letters (10) to Sir RI Murchison 1850-55. *Edinburgh Univ Libr* (MSS Gen 523/4, 6)

See also Liversidge

CLERK-MAXWELL, see Maxwell

[122] **CLIFT, William** (1775-1849)
FRS, naturalist

Correspondence and papers 1791-1842, incl catalogues, accounts, reports and diaries rel to the Hunterian Museum, accounts of his early life and apprenticeship, natural history MSS and notes, and his transcripts of works by J Hunter. *R Coll of Surgeons, London*
Scientific correspondence (27 vols) of Clift and of Sir R Owen, memoranda rel to specimens sold to the Royal College of Surgeons of England, index to Sir E Home's natural history MSS, and other papers. *Brit Mus (Nat Hist) General Libr*
Personal and family correspondence 1792-1849. *Brit Libr* (in Add MSS 39954-55)
Original drawings and copper plates of engravings to illustrate M Baillie's *Morbid anatomy* 1793-1802. *R Coll of Physicians, London*

See also Home, Owen

[123] **COCKCROFT, Sir John Douglas,** OM, KCB (1897-1967)
FRS, physicist

Papers and correspondence (127 boxes) 1921-67, incl research notebooks, lecture scripts and notes, college administration, committee and conference papers, and MSS of his published works. *Churchill Coll, Cambridge* [NRA 14614]
Private office papers as director of the Atomic Energy Research Establishment, Harwell (1946-58). *Public Record Office* (in AB 27)
Papers rel to the Maud Committee 1940-41, and correspondence with Sir J Chadwick 1940-60. *Atomic Energy Research Establishment, Harwell*
Correspondence (53 items) with N Bohr 1930-61. *Niels Bohr Inst, Copenhagen*
Correspondence (*c*13 items) with EC Stoner 1931-66. *Brotherton Libr, Leeds Univ* [NRA 17735]
Correspondence with Sir RE Peierls 1939-66. *Bodleian Libr, Oxford* [NRA 20805]
Correspondence with Lord Blackett 1939-66. *Royal Soc* [NRA 22627]
Correspondence with JR Oppenheimer 1940-67 and with G Garnow. *Libr of Congress, Washington*
Correspondence with Viscount Cherwell 1943-57. *Nuffield Coll, Oxford* [NRA 16447]

[124] **COLEY, Henry** (1633- ?1695)
Mathematician

Papers, incl 'Principles of geometry', astrological and mathematical collections 1665-95, and geometrical and astrological fragments from Euclid. *Brit Libr* (Sloane MSS 1405, 2279-85, 2328, 3880)

[125] **COLLIN, James Edward** (1876-1968)
Entomologist

Entomological notes and correspondence. *Hope Libr, University Mus, Oxford*

[126] **COLLINS, John** (1625-1683)
FRS, mathematician

Mathematical papers and correspondence (50 items) collected by Sir I Newton for publication in part in *Commercium epistolicum* (1712). *Royal Soc*
Mathematical papers; correspondence with J Pell 1666-69. *Brit Libr* (in Sloane MSS 2279, 2281; Add MSS 4278, 4280, 4294, 4394, 4414, 4432)
Transcript of Sir I Newton's 'De analysi'; correspondence (57 items) with J Gregory 1668-75 and others. *St Andrews Univ Libr*
Correspondence (over 200 items) with I Barrow, J Flamsteed, J Gregory, Sir I Newton, W Oughtred and others 1663-78. *Private*
Correspondence with Sir I Newton 1671-83. *Cambridge Univ Libr* (MS Add 3977)

▷SP Rigaud, *Correspondence of scientific men of the seventeenth century*, 2 vols, 1841-62

See also Barrow, Flamsteed, Gregory J (1638-1675), Wallis

[127] **COLLINSON, Peter** (1694-1768)
FRS, naturalist

Commonplace books (2) containing some correspondence *c*1708-68; letters (36) to C Linnaeus 1739-67. *Linnean Soc*
Letters addressed to him 1725-68. *Brit Libr* (Add MSS 28726-27)
Remarks on the Revd R Forster's dissertation on swallows 1762. *Ibid* (in Add MS 4440)
Papers rel to his association with Lord Petre and their horticultural interests incl notes on rare trees planted *c*1730-60. *Essex RO, Chelmsford*
Account of the introduction of American seeds into Britain 1766. *Brit Mus (Nat Hist) Botany Dept*
Correspondence (92 items) with C Colden, 1741-68, and others. *New York Hist Soc*
Correspondence (*c*55 items) 1732-68, principally with B Franklin 1750-63. *American Philosophical Soc, Philadelphia*
Correspondence (27 items) with Sir H Sloane 1728-42. *Brit Libr* (in Sloane MSS 1968, 3322, 4025, 4053-55, 4057-58, 4069)
Correspondence (18 items) with EM da Costa 1741-66. *Ibid* (in Add MS 28536)
Letters (33) to A von Haller. *Berne Public Libr, Switzerland*
Letters, incl 32 to E Robinson. *Soc of Friends Libr, London*
Letters (23) to J Custis 1734-46. *American Antiquarian Soc, Worcester, Mass*
Letters (16) from J Custis 1734-42. *Libr of Congress, Washington*
Letters (15) to T Birch 1753-65. *Brit Libr* (in Add MS 4303)
Miscellaneous letters and papers (14). *Royal Soc*

See also Lambert

[128] **COOKE, Sir William Fothergill** (1806-1879)
Electrician

Correspondence and papers (7 vols) 1834-79, incl personal and business correspondence and papers rel to the arbitration between him and Sir C Wheatstone. *Inst of Electrical Engineers* [NRA 20573]
Papers on the construction of the Paddington-Slough telegraphs. *Public Record Office* (British Transport Historical Records)

[129] **COOPER, Sir Astley Paston,** 1st Bt (1768-1841)
FRS, surgeon

Correspondence and papers, incl notes for anatomical and surgical lectures 1796-1827, notebooks, case notes, post-mortem reports, account book at Paris 1792, and MSS of published works. *R Coll of Surgeons, London* [NRA 9521]
Lectures on surgery 1806-07. *R Coll of Physicians, Edinburgh*
Notes on the Anatomy Bill 1832. *Osler Libr, McGill Univ, Montreal*
Correspondence (10 items) with Sir R Peel 1823-26. *Brit Libr* (in Add MSS 40359, 40366, 40370-71, 40377-78, 40390)

[130] **COPELAND, Ralph** (1837-1905)
Astronomer

Correspondence and papers 1874-1905, incl correspondence, observations, calculations and diaries rel to Dunecht Observatory, papers rel to Mauritius, eclipses, meteorology and seismology, and administrative records as astronomer royal for Scotland. *R Observatory, Edinburgh*
Letters (128) to officers of the Royal Astronomical Society 1873-99. *R Astronomical Soc*
Letters (37) to Sir D Gill 1874-79. *R Geographical Soc*

COSTA, see Da Costa

[131] **COUCH, Jonathan** (1789-1870)
Naturalist

Papers incl a journal (12 vols) 1805-70, materials for a history of British crustacea, and a natural history of Cornish fishes. *Linnean Soc*
Natural history journal (2 vols) 1866. *Cheltenham Public Libr* (Francis Day Papers)
Notes on natural history and an album of watercolours (243) for his *History of the fishes of the British islands* (1860-65). *Royal Inst of Cornwall, Truro*
A natural history of Cornish fishes and a history of Polperro. *Redruth Public Libr*
A natural history of Cornish fishes 1822. *Blacker-Wood Libr, McGill Univ, Montreal*
Drawings (47) of primitive marine life etc; notes on books on fishes by other authors. *Brit Mus (Nat Hist) Zoology Dept*
Letters to ACLG Günther 1860-67. *Ibid, General Libr*

[132] **COWPER, William** (1666-1709)
FRS, surgeon

Papers 1681-1703, incl medical discourses, prescriptions, observations, case notes, descriptions of surgical operations with drawings, herbal indexes, botanical notes and drawings, letters (10) to Sir H Sloane. *Brit Libr* (in Sloane MSS 3323, 3408-09, 4039, 4058, 4078)

[133] **CROMBIE, Revd James Morrison** (1833-1906)
Botanist

MS (5 vols) of his *Monograph of lichens found in Britain*, Part I (1894), with unpublished introduction (1 vol), draft of Part II (2 vols) and enumeration of lichens (3 vols). *Brit Mus (Nat Hist) Botany Dept*

[134] **CROMPTON, Rookes Evelyn Bell** (1845-1940)
FRS, electrical and transport engineer

Business records of Crompton-Parkinson Ltd, incl Crompton's notebooks with calculations (3 vols), diaries and account books 1855-1933, letter books (4 vols) 1870-1920, miscellaneous correspondence and papers (1 file), press cuttings (3 vols) 1886-1935. *Science Mus Libr*

[135] **CROOKES, Sir William,** OM (1832-1919)
FRS, chemist

Laboratory notebooks (16) 1881-1919; letters (9). *Royal Inst* [NRA 9522]
Notebooks (7) 1858-83; laboratory weighing book 1860; spectral photography records 1900-15; letters to CH Gimingham 1871-77. *Science Mus Libr* [NRA 9524]
Correspondence (168 items) with Sir GG Stokes 1862-99. *Cambridge Univ Libr* (in MS Add 7656)
Correspondence (56 items) with Sir OJ Lodge 1885-1916. *Incorporated Soc for Psychical Research* [NRA 11857]
Letters (50), incl 10 to Sir JFW Herschel 1858-63. *Royal Soc*
Letters (24) to HE Armstrong 1885-1915. *Imperial Coll, London* [NRA 11420]
Letters (22) to SP Thompson 1876-1916. *Ibid* [NRA 11421]

[136] **CULLEN, William** (1710-1790)
FRS, physician

Correspondence and papers, incl letters (41 vols) addressed to him 1755-90, a letter book 1782, notes (1 vol) of medicines supplied by him 1737-41, his own and student notes (29 sets) of his lectures on medicine, physiology and chemistry *c*1755-78. *R Coll of Physicians, Edinburgh* [NRA 16015]
Papers, incl lecture notes on botany and agriculture, directions for collecting and preserving natural curiosities, notes on the classification of birds and insects, and botanical notes. *Glasgow Univ Libr*
Clinical case papers and reports *c*1763-72, notes of his lectures on the practice of physic 1769-71. *R Coll of Physicians, London*
Notes (9 vols) of his lectures on medicine. *Bodleian Libr, Oxford* (MSS Eng misc d 298-306)
Notebook of female clinical case histories 1784, and student notes (6 sets) of his lectures 1766-87. *Edinburgh Univ Libr* (in MSS Dc 1, 6, 7, 10; MS Dk 3, 5)

See also Hope TC

[137] **CURTIS, John** (1791-1862)
Entomologist

Entomological diary 1840-54 and some letters. *R Entomological Soc*
Water-colour drawings (770) for his *British entomology* (1824-39). *Brit Mus (Nat Hist) Entomology Dept*
Pencil drawings (with C Curtis) for *Plantae Javanicae rariores* by T Horsfield, JJ Bennett and R Brown (1838-52). *Ibid, Botany Dept*
Notes and drawings. *Hope Libr, University Mus, Oxford*

[138] **CURTIS, William** (1746-1799)
Botanist

Correspondence and papers, incl lectures, notes and MSS on botany and entomology. *Curtis Mus, Alton*
Observations on the sand wasp, larder beetle

1788, and a method of taking wild duck 1790. *Linnean Soc*
Drawings for plates to his *Botanical Magazine*, Vols 1-18 (1787-1804). *R Botanic Gardens, Kew*

[139] **DA COSTA, Emanuel Mendes** (1717-1791)
FRS, naturalist

Illustrated catalogue (6 vols) of fossils in his collection, part II of an 'Index fossilium', and 'Notary business' 1762-68. *R Coll of Surgeons, London*
Catalogue of his library 1781. *Brit Libr* (Add MS 9389)
Papers (12 items) submitted to the Royal Society. *Royal Soc*
Notes and extracts principally about the Jews, incl his own family. *Brit Libr* (Add MSS 29867-68)
Correspondence (11 vols) 1737-87. *Ibid* (Add MSS 28534-44)
Correspondence (200 items) with T Pennant and Lord Penrhyn 1752-88. *Warwick County RO* (TP 408) [NRA 23685]
Correspondence and papers (*c*60 items) largely rel to Derbyshire minerals *c*1748-67. *Derby Central Libr*
Correspondence and papers (19 items) 1749-76. *Fitzwilliam Mus, Cambridge* (SG Perceval Colln)

See also Collinson, Fothergill

[140] **DALE, Sir Henry Hallett,** OM, GBE (1875-1968)
FRS, physician

Papers (60 boxes). *Royal Soc*
Papers 1888-1964, incl notes (6 vols) of experiments on animals 1913-16, lectures and articles. *Contemporary Medical Archives Centre, Wellcome Inst for the Hist of Medicine* [NRA 24904]
Correspondence (*c*25 items) with Sir T Lewis 1921-42. *Ibid* [NRA 24907]
Correspondence and papers rel to research topics and administration. *Medical Research Council*
Correspondence as secretary of the Royal Society with E Hindle 1925-28. *Glasgow Univ Archives* [NRA 21908]
Correspondence (over 20 items) with AV Hill. *Churchill Coll, Cambridge*

[141] **DALE, Samuel** (1659-1739)
Physician

Journal of his journeys between Braintree and Cambridge 1722-38. *Cambridge Univ Libr*
Catalogue of J Ray's 'Hortus siccus'; annotated copies of T Johnson's *Iter plantarum investigationis* (1629, 1632) and J Ray's *Synopsis methodicum stirpium Britannicarum* (1690). *Brit Mus (Nat Hist) Botany Dept*
MS life of J Ray. *Bodleian Libr, Oxford* (MS Rawl Essex 21)
Correspondence (*c*50 items) with Sir H Sloane 1692-1736, and letters (*c*20) to J Petiver 1700-17. *Brit Libr* (Sloane MSS 3321-4068 *passim*)
Letters (60) to W Holman *c*1711-28. *Essex RO, Chelmsford*

[142] **DALTON, John** (1766-1844)
FRS, chemist

Papers and correspondence, incl meteorological observations (9 vols) 1787-92, 1803-27, notes for lectures on astronomy, meteorology, physics, chemistry, optics, acoustics and natural philosophy *c*1791-1834, papers (12) read to the Manchester Literary and Philosophical Society 1801-36, MSS of his published works, account books (14) 1792-1838, and correspondence (67 items) 1788-1844. *John Rylands Univ Libr, Manchester*
Herbarium (2 vols) 1791-93. *R Botanic Garden, Edinburgh*
Meteorological register (2 vols) 1803-27. *Science Mus Libr*

See also Sowerby

[143] **DANIELL, John Frederic** (1790-1845)
FRS, physicist

Letters to him from other scientists. *King's Coll, London*
Letters (23) to the Society for the Diffusion of Useful Knowledge 1828-38. *University Coll London*
Letters (10) to Sir JW Lubbock 1833-42. *Royal Soc*

[144] **DARWIN, Sir Charles Galton,** KBE (1887-1962)
FRS, physicist

Correspondence (40 items) with N Bohr 1913-45. *Niels Bohr Inst, Copenhagen*
Correspondence with the Eugenics Society 1930-60. *Contemporary Medical Archives Centre, Wellcome Inst for the Hist of Medicine* [NRA 24905]
Miscellaneous correspondence. *Medical Research Council*

[145] **DARWIN, Charles Robert** (1809-1882)
FRS, naturalist

Papers and correspondence (157 vols) 1825-82, incl notes on lectures attended and books read, diary 1826, notes and observations made during the voyage of HMS *Beagle* 1832-36, notes for and MSS of his published works, and accounts with his publisher 1881. *Cambridge Univ Libr*

Papers and correspondence, incl 18 notebooks kept during the voyage of HMS *Beagle* 1831-36, specimen and other notebooks, and letters written to his sisters during the voyage. *Down House, Downe* [NRA 22849]

Lists and notes on reptiles and amphibians collected during the voyage of HMS *Beagle* 1831-36. *Brit Mus (Nat Hist) Zoology Dept*

Correspondence, incl *c*450 letters to Sir C Lyell 1837-74 and 87 to GJ Romanes 1874-82. *American Philosophical Soc, Philadelphia*

Letters (176) to TH Huxley 1851-82. *Imperial Coll, London*

Letters (155) to WD Fox 1828-78, and an extract from his *Insectivorous plants* (1875). *Christ's Coll, Cambridge*

Letters (155) to A Gray 1855-81. *Houghton Libr, Harvard Univ, Cambridge, Mass*

Correspondence (153 items) with AR Wallace 1857-81. *Brit Libr* (Add MS 46434)

Correspondence (136 items) with W Tegetmeier 1855-81, and 12 letters to A Hancock 1849-54. *New York Botanical Garden Libr*

Correspondence, incl copies of his letters (3 vols) to Sir JD Hooker 1843-66, correspondence (45 items) with JS Henslow 1831-37 and letters to Sir WT Thiselton-Dyer 1873-81. *R Botanic Gardens, Kew*

Letters (65) 1858-82, incl 46 to E Krause. *Huntington Libr, San Marino*

Letters (64) *c*1857-81, incl 41 to Lord Farrer 1868-81. *Linnean Soc*

Letters (36) to HW Bates 1860-82, and 27 to the Revd JB Innes 1846-81. *Case Western Reserve Univ, Cleveland, Ohio* (Robert M Stecher Colln)

Letters (44) to Sir F Galton. *University Coll London*

Correspondence (38 items) with Sir JS Burdon-Sanderson. *Woodward Biomedical Libr, Univ of British Columbia, Vancouver* (Sinclair Colln)

Letters (35) to J Tyndall. *Royal Inst*

Letters (30) to ACLG Günther 1860-81, and miscellaneous letters. *Shrewsbury School*

Letters (16) to R Fitch *c*1850. *Castle Mus, Norwich* [NRA 9528]

Letters (13) to the Royal Geographical Society 1839-74. *R Geographical Soc*

Letters (11) to Sir JW Lubbock 1846-56. *Royal Soc*

Letters (11) to A Newton 1863-77. *Balfour and Newton Libr, Zoology Dept, Cambridge*

Miscellaneous letters and transcripts (71 items) 1844-80. *Brit Mus (Nat Hist) General Libr*

Letters (*c*40) from him concerning the Darwin family estates in Lincolnshire *c*1846-62. *Lincolnshire Archives Office* (in HIG 4) [NRA 18982]

▷*Handlist of Darwin papers at the University Library Cambridge*, 1960; J Browne, 'The Charles Darwin-Joseph Hooker correspondence: an analysis of manuscript resources and their use in biography', *J Soc Bibliog Nat Hist*, viii, pt 4, 1978, pp351-66

See also Bates, Burdon-Sanderson, Darwin F, Darwin GH, Galton, Günther ACLG, Henslow, Hooker JD, Huxley TH, Owen, Thiselton-Dyer

[146] **DARWIN, Sir Francis** (1848-1925)
FRS, botanist

Family and general correspondence and papers *c*1866-1924, incl speeches and articles. *Brit Libr* (Add MSS 58375-80, 58394-97)

Miscellaneous correspondence and papers, incl material rel to his editions of his father's letters. *Cambridge Univ Libr*

Letters (35) to TH Huxley 1882-90. *Imperial Coll, London*

Miscellaneous correspondence (27 items) 1878-1912. *American Philosophical Soc, Philadelphia*

Correspondence (12 items) with FO Bower 1913-24. *Glasgow Univ Archives* [NRA 18906]

Letters (12) to Sir A Geikie 1885-87. *Edinburgh Univ Libr* (in MS Gen 524)

Letters to Sir F Galton *c*1885-1909. *University Coll London*

[147] **DARWIN, Sir George Howard,** KCB (1845-1912)
FRS, mathematician and astronomer

Correspondence and papers, incl a diary of the British Association meeting in South Africa 1905, papers on tides 1882-84, mathematical computations 1882-84, notes on climbing plants, notes taken by him of lectures by EJ Routh on physics, astronomy etc, and letters to his parents 1863-82. *Cambridge Univ Libr*

Correspondence 1899-1912, and letters and papers rel to the International Geodetic Conference 1909. *R Geographical Soc*

Correspondence and papers rel to the changes in Cambridge University 1897-1904. *Cambridge Univ Archives*

Correspondence (139 items) with Lord and Lady Kelvin 1878-1907. *Glasgow Univ Libr*

Correspondence (65 items) with Lord Kelvin 1878-1906. *Cambridge Univ Libr* (in MS Add 7342)

Correspondence with Sir F Galton 1870-1910. *University Coll London* [NRA 19968]

Letters (51) to Sir D Gill 1900-12. *R Geographical Soc*

Letters (49) to officers of the Royal Astronomical Society 1883-1900, and a report on the total solar eclipse of 1870. *R Astronomical Soc*

Letters (21) to Sir GG Stokes 1876-1902. *Cambridge Univ Libr* (in MS Add 7656)
Letters (10) to O Browning 1882-1906. *Hastings Public Libr*

[148] **DAUBENY, Charles Giles Bridle** (1795-1867)
FRS, chemist and botanist

Papers 1810-67, incl commonplace books (2 vols), notes taken by him at lectures (3 vols), notes for his lectures (30 vols), and on chemistry (10 vols), geology (2 vols), botany (2 vols), agriculture and rural economy (2 vols), and on experiments (3 vols), meteorological register 1861-67, and 120 letters from scientists. *Magdalen Coll, Oxford* (MSS 375, 377-94, 400)
Papers, incl lectures on rural economy and on botany, diary 1834-67, notebook of botanical excursions 1839-65, notes on the genera of plants and on soil analysis, unpublished biography of J Sibthorp 1851, annotated incomplete copies (2) of his *Oxford Botanic Garden* (1864), notes on European agricultural schools, catalogues, and correspondence with HG Bohn 1845-46. *Bodleian Libr, Oxford* (MSS Sherard 248, 263-398) [NRA 6305]
Letters (55) to Sir WJ Hooker 1837-64. *R Botanic Gardens, Kew*
Letters (20) to Sir WC Trevelyan *c*1845-1864. *Newcastle Univ Libr* [NRA 12238]
Letters (15) to the Revd T Burgess 1809-25. *Bodleian Libr, Oxford* (in MS Eng lett c 134)

[149] **DAVIDSON, Thomas** (1817-1885)
FRS, palaeontologist

Papers and drawings on brachiopods (22 vols). *Brit Mus (Nat Hist) Palaeontology Dept*
Illustrated paper on the geology of the Paris basin 1840. *Geological Soc of London*
Letters, incl 66 to G Maw about Silurian brachiopods *c*1880-83. *Inst of Geological Sciences* [NRA 18675]
Letters (13) to Sir C Lyell 1852-70. *Edinburgh Univ Libr* (in MS Lyell 1)
Letters (13) to Sir A Geikie 1867-79. *Ibid* (in MS Gen 524)
Letters to R Fitch about E Anglian geology. *Castle Mus, Norwich*

[150] **DAVIES, Revd Hugh** (*c*1731-1821)
Naturalist

Correspondence and papers, incl 116 letters addressed to him, draft letters and notes, and botanical miscellanea. *Nat Libr of Wales* (MSS 6664-65)
Notes and drawings of British fungi (1 vol). *Brit Mus (Nat Hist) Botany Dept*
Letters to him and copies made by him of S Brewer's botanical diary kept in N Wales 1726-27 and R Richardson's directions to JJ Dillenius about Welsh plants. *Nat Mus of Wales Botany Dept, Cardiff*
Letters (34) to Sir JE Smith 1790-1820. *Linnean Soc*

See also Sowerby

[151] **DAVY, Sir Humphry,** Bt (1778-1829)
FRS, chemist

Papers and correspondence, incl 5 laboratory notebooks partly written in collaboration with Faraday 1805-29, papers (2 vols) mainly on the safety lamp, notes for his lectures on geology, agriculture, the history of science etc 1802-12, literary and scientific notebooks 1795-1829, annotated copies of his published works, and family and scientific correspondence (*c*350 items). *Royal Inst*
Transcripts of his letters, medical and philosophical papers and poems, with biographical notes, made by his brother and sister-in-law 1830-31 (6 vols). *Keele Univ Libr* (Raymond Richards Colln) [NRA 1085]
Drafts for papers published in *Philosophical Transactions*, referee reports, official correspondence as President of the Royal Society and other papers. *Royal Soc*
Notes for lectures, and working papers. *R Geological Soc of Cornwall, Penzance*
Reports and correspondence rel to his 'protectors' for ships' hulls 1823-25. *Nat Maritime Mus, Greenwich* (MSS Adm BP/43A, 44A, 45) [NRA 9518]
Letters (69) to his family 1799-1829. *Science Mus Libr* (MS 333) [NRA 9524]
Letters (26) to AJG Marcet 1803-22. *American Philosophical Soc, Philadelphia*
Correspondence (23 items) with JJ Berzelius 1808-25. *R Swedish Academy of Sciences, Stockholm*
Letters (23) to JG Children 1807-24. *Brit Libr* (Add MS 38625)
Correspondence (20 items) with Sir R Peel 1823-26. *Ibid* (Add MSS 40356-89 *passim*)
Letters (12) to M Faraday 1815-23. *Inst of Electrical Engineers* (Blaikley Colln)
Letters to J Buddle and others on the safety lamp 1816-30, with some correspondence (9 items) concerning a presentation made to him 1816-17. *Private* [NRA 11184]

See also Brodie BC the younger, Faraday

[152] **DAWKINS, Sir William Boyd** (1837-1929)
Geologist and palaeontologist

Correspondence 1868-1928 and papers, incl notes on Pleistocene mammals 1872, report on the Dover coalfield 1891, and MSS of articles and lectures. *Derbyshire RO, Matlock*

Diary of his exploration of Wookey Hole, and notes on quaternary palaeontology. *Wells Mus, Somerset*
MSS on caves and cave faunas *c*1860-69. *University Mus, Oxford* (Geological Collns)
Letters (252) from EA Freeman 1866-92, with *c*100 from JR Green 1859-71. *Jesus Coll, Oxford* (MSS 192-93, 198-200)
Miscellaneous letters addressed to him 1880-1917. *Bodleian Libr, Oxford* (in MS Eng lett d 169)

[153] **DEANE, Sir Anthony** (?1638-1721)
FRS, shipbuilder

Letters (31) to Samuel Pepys 1666-89, with other papers incl observations on the war of 1665-67 and on naval and shipbuilding affairs 1671-85. *Bodleian Libr, Oxford* (in MSS Rawl A 170-464 *passim*)
'Doctrine of naval architecture' 1670, description of a method of calculating a ship's draught, and observations on the improvement of frigates in sailing 1685. *Magdalene Coll, Cambridge* (MSS 2501, 2872, 2910)
Letters and papers (15 items), incl letters to Sir R Clayton 1662-79. *Nat Maritime Mus, Greenwich* (LBK 3)

[154] **DEE, Revd John** (1527-1608)
Mathematician

Correspondence and papers, incl mathematical works, account of discoveries 1577, directions for a voyage to Cathay 1580, 'Mysteriorum libri' 1583-1607, autobiography 1592, corrections to Sigbert's chronicle, theological correspondence with R Edwards. *Brit Libr* (in Cotton MSS Vitell C VII, IX, Otho E VIII, Aug I.i, Append XLVI; Cotton Ch XIV 1)
Letters and papers, incl 'Epilogismus calculi diurnus planetarum', treatise on Rosicrucian secrets, paper on maritime jurisdiction. *Ibid* (in Harleian MSS 249, 374, 532, 6485, 6986)
Letters and papers, incl his 'Histories of truth and true philosophy' 1574, directions for a voyage to Cathay 1580. *Ibid* (in Lansdowne MSS XIX, LXI, CXXII)
'Thalattokratia Bretannike' 1597. *Ibid* (in Royal MS 7 C XVI)
'Mysteriorum libri' 1581-83, 'Claves angelicae' 1584-85, instructions and annotations on Euclid's *Elements* 1569. *Ibid* (in Sloane MSS 15, 3188, 3191)
Diaries 1577-1601, papers on navigation 1576 and the reform of the calendar 1583, notes and papers on arithmetic, astrology, alchemy, and fen drainage. *Bodleian Libr, Oxford* (in MSS Ashmole 174, 242, 337, 487-88, 1394, 1426, 1451, 1503, 1789)
Chemical diary 1581. *Ibid* (MS Rawl D 241)
Autobiography 1592. *Ibid* (MS Smith 96)
Commentary on a work of Ptolemy 1550. *Ibid* (MS Arch Selden B 8)
Notes on a nocturnal visitation etc 1597. *Ibid* (MS Bodl 485)
Notes in a theological commonplace book. *Ibid* (MS Dugdale 24)
Alchemical notes. *Ibid* (MS e Mus 63)
Literary miscellanies. *Ibid* (MS Douce 363)
Tract on the rules of exchange of moneys 1578. *Harvard Univ Business School, Boston, Mass*
Catalogues of his books and manuscripts. *Brit Libr* (Harleian MS 1879, Add MS 35213), *Bodleian Libr* (CCC MS 191), *Trinity Coll, Cambridge* (MS 0.4.20)

▷R Deacon, *John Dee*, 1968

DE FERRANTI, see Ferranti

[155] **DE LA BECHE, Sir Henry Thomas** (1796-1855)
FRS, geologist

Correspondence (27 vols) and papers, incl travel diaries 1818-29, notes, MSS of geological publications, draft report on the Geological Survey and the Museum of Economic Geology 1841. *Nat Mus of Wales Geology Dept, Cardiff*
Correspondence 1823-55 and papers, incl travel diaries 1816-18, notebooks (2) on the geology of the West Country and South Wales 1830-40, drawings for and annotated copies of his published works, official correspondence and papers as director general of the Geological Survey 1835-55. *Inst of Geological Sciences* [NRA 18675]
Notebook containing geological sketches made in South Wales *c*1840. *Glamorgan Archive Service, Cardiff* (Nicholl of Merthyr Mawr Papers) [NRA 9673]

See also Buckland, Logan, Lyell, Murchison, Owen, Phillips J, Playfair, Ramsay AC

[156] **DE LA RUE, Warren** (1815-1889)
FRS, chemist and astronomer

Letters (128) to officers of the Royal Astronomical Society 1851-88, 2 papers submitted to the Society, correspondence (52 items) with Sir JFW Herschel 1856-71, and letters (11) to R Hodgson 1861-67. *R Astronomical Soc*
Scientific papers, incl material on solar physics in collaboration with B Stewart and B Loewy and papers on sunspots. *Royal Soc*
Correspondence (235 items) with Sir GG Stokes 1855-85. *Cambridge Univ Libr* (in MS Add 7656)
Correspondence (18 items) with Sir GB Airy 1855-72. *R Greenwich Observatory, Herstmonceux* (in MSS 946-52)
Letters (11) to M Faraday 1849-61. *Inst of Electrical Engineers* (Blaikley Colln)

Letters from him about his properties in Guernsey and other papers *c*1860-80. *Reading Univ Libr*

DE MAYERNE, see Mayerne

[157] **DE MORGAN, Augustus** (1806-1871)
Mathematician

Papers and correspondence incl 20 mathematical notebooks *c*1822-30, 200 notebooks rel to his teaching of mathematics and logic *c*1843-66, notes for and drafts of published works 1845-68, and correspondence (*c*300 items) 1828-70. *London Univ Libr* [NRA 22570]

Papers and correspondence incl notes for lectures 1828, 1847, 1862, mathematical papers 1849-51, and correspondence (over 270 items) 1827-68. *University Coll London* (in MSS Add 2-3, 5-7, 97; College Correspondence; SDUK Papers; Brougham Papers; London Mathematical Society Papers) [NRA 14258, 22241]

Papers rel to the Royal Astronomical Society's standard scales 1833-36, annotated copies of his published works, 34 letters to the Society's officers 1831-66, and over 150 letters addressed to him 1835-63. *R Astronomical Soc*

Catalogue of his mathematical books. *R Observatory, Edinburgh* [NRA 9504]

Correspondence (*c*384 items) with Sir JFW Herschel 1831-70; letters (26) to Sir JW Lubbock 1832-60. *Royal Soc*

Letters (252) to Sir WR Hamilton 1841-65. *Trinity Coll, Dublin* (MS 1493) [NRA 20076]

Correspondence (205 items) with Sir GB Airy 1850-70. *R Greenwich Observatory, Herstmonceux* (in MSS 942-52)

Letters (*c*45) to C Babbage 1830-50. *Brit Libr* (in Add MSS 37185-94, 37200)

Letters (17) to Sir GG Stokes 1847-64. *Cambridge Univ Libr* (in MS Add 7656)

Letters (12) to JO Halliwell-Phillipps 1838-67. *Edinburgh Univ Libr*

Letters (10) to Lord Kelvin 1845-49. *Cambridge Univ Libr* (in MS Add 7342)

See also Babbage, Boole, Horner, Sheepshanks

[158] **DESAGULIERS, Revd John Theophilus** (1683-1744)
FRS, natural philosopher

Papers 1729-41, incl accounts of electrical and other experiments, dissertations on strength and on bodies in motion, account of Bianchini's discoveries concerning Venus, description of his machine to show tidal phenomena, and notes on elasticity. *Brit Libr* (Add MSS 4432-37)

Mathematical notebooks (2), 1709 and 1712. *Scientific Periodicals Libr, Cambridge*

Papers (59) submitted to the Royal Society. *Royal Soc*

[159] **DESCH, Cecil Henry** (1874-1958)
FRS, metallurgist

Correspondence (*c*1,600 items) and papers *c*1885-1957, incl diaries, notebooks and other working papers, lecture notes, texts of articles, committee papers. *Sheffield Univ Libr* [NRA 17354]

Notebooks (2) incl analyses of ancient metals 1929, 1937, and some correspondence. *Brit Mus Research Laboratory*

[160] **DEWAR, Sir James** (1842-1923)
FRS, chemist

Correspondence and papers, incl laboratory and other notebooks, lecture notes 1878-91, papers rel to the St Louis International Exhibition 1904 and his World War I work. *Royal Inst* [NRA 9522]

Letters (68) to HE Armstrong 1890-1920. *Imperial Coll, London* [NRA 11420]

Correspondence (33 items) with Lord Kelvin 1883-1907. *Cambridge Univ Libr* (in MS Add 7342)

Letters (29) to Sir GG Stokes 1875-1908. *Ibid* (in MS Add 7656)

Letters (17) to Lord Rutherford 1906-22. *Ibid* (in MS Add 7653)

Letters (17) to Sir JJ Thomson. *Ibid* (in MS Add 7654)

Letters (13) to Lord Playfair 1885-98. *Imperial Coll, London* [NRA 11556]

Letters (10) to Sir D Gill 1907-12. *R Geographical Soc*

Correspondence with GE Hale 1909-23. *California Inst of Technology, Pasadena*

[161] **DILLENIUS, Johann Jakob** (1687-1747)
FRS, botanist

Correspondence 1726-46 and papers, incl botanical descriptions, notes and drawings, additions to plant lists by J Bobart and P Hermann, notes on the Petiver Museum, annotated abridgment of his *Historia muscorum* 1740, correspondence (16 items) with A von Haller 1738-46, and letters to R Richardson. *Bodleian Libr, Oxford* (MSS Sherard 30, 33, 182, 202-10; MSS Radcliffe Trust c 1-11) [NRA 6305]

'Fungi catalogi Gissensis' with drawings, 17 letters to S Brewer 1726-28, drawings for first 79 plates of *Historia muscorum* (1741), and other botanical drawings. *Brit Mus (Nat Hist) Botany Dept*

Transcript of his notes on plants observed in N Wales 1726. *Nat Mus of Wales Botany Dept, Cardiff*

Letters (11) to C Linnaeus 1739-46. *Linnean Soc*

See also Davies

[162] **DIXEY, Frederick Augustus** (1855-1935)
FRS, physiologist

Papers and correspondence, incl diary (21 vols) 1868-1935, 3 papers on applied logic 1881, MS 'On Aristotle as a biologist' *c*1923, and his lists of his own publications. *Bodleian Libr, Oxford* [NRA 2654]
Entomological papers, observations, and correspondence. *Hope Libr, University Mus, Oxford*
Clerical correspondence, religious writings etc 1888-1934. *Pusey House Libr, St Cross Coll, Oxford* [NRA 2654, 19807]

[163] **DIXON, Harold Baily** (1852-1930)
FRS, chemist

Notebook (*c*1892) and 6 letters addressed to him 1891-94. *John Rylands Univ Libr, Manchester* [NRA 9194]
Correspondence (16 items) with A Smithells 1891-1927. *Brotherton Libr, Leeds Univ* [NRA 20283]

[164] **DOBELL, Clifford** (1886-1949)
FRS, zoologist

Papers 1899-1943, incl drawings, plates and MSS of published and unpublished works. *Wellcome Hist Medical Libr* (MSS 2197-2220)
Miscellaneous correspondence, mainly rel to bacteriology and virology. *Medical Research Council*
Correspondence with E Hindle 1925-26. *Glasgow Univ Archives* [NRA 21908]
See also Bateson

[165] **DONKIN, Bryan** (1768-1855)
FRS, engineer

Diaries 1816-17, and correspondence. *Private*
Contribution to a volume of 'Original entries of the comparisons of various standard scales' 1834-36; 3 letters to officers of the Royal Astronomical Society 1824-40. *R Astronomical Society*
Correspondence (20 items) with C Babbage 1829-38. *Brit Libr* (in Add MSS 37184-88, 37190-91)

[166] **DONNAN, Frederick George** (1870-1956)
FRS, chemist

Correspondence and papers 1906-56, incl reports, speeches, obituaries and papers on the Scientific Relief Fund and its committee. *University Coll London* [NRA 14720]
Personal papers. *Private* [NRA 16920]
Letters to the Vice-Chancellor of Liverpool University 1907-11. *Liverpool Univ Libr* [NRA 16920]

[167] **DOUGLAS, James** (1675-1742)
FRS, physician

Papers and correspondence, incl lecture notes, catalogue of his medical books at Utrecht 1698, other lists, catalogues and bibliographies, MSS of his *Description of the peritonaeum* (1730) and his *English pronunciation* (1956), unpublished lives of J Ray and of Horace, notes for his projected book on osteology, medical case notes 1700-12, accounts of dissections, anatomical drawings, and other papers rel to botany, zoology, anatomy, phonetics and grammar. *Hunterian Libr, Glasgow Univ*
Medical, botanical and zoological papers submitted to the Royal Society. *Royal Soc*
▷KB Thomas, *James Douglas of the pouch and his pupil William Hunter*, 1964, and CH Brock, 'The rediscovery of James Douglas', *The Bibliotheck*, viii, 1976-77, pp168-76

[168] **DRUCE, George Claridge** (1850-1932)
FRS, botanist

Correspondence *c*1870-1932 and papers, incl plant lists and annotated copies of botanical publications by himself and others. *Botany Dept, Oxford*
Scrapbook of botanical and other correspondence and papers *c*1882-1902. *History of Science Mus, Oxford* (MS Mus 110)
Miscellaneous papers and drawings collected by him, with some related papers of his own. *Bodleian Libr, Oxford* (MSS Sherard 407-48) [NRA 6305]

[169] **DRURY, Dru** (1725-1803)
Naturalist

Scientific, family and personal papers, incl letter book 1761-83 and notebook rel to his *Illustrations of natural history* (1770-82). *Brit Mus (Nat Hist) Entomology Dept*
Entomological notebooks. *Hope Libr, University Mus, Oxford*

[170] **DRYANDER, Jonas** (1748-1810)
Botanist

Correspondence and papers, incl list of plants collected by F Masson in Madeira, catalogue of plants in the Jardin Royal des Plantes, Paris 1777, notes on the ericaceae, and 51 letters from CL L'Héritier de Brutelle 1785-90. *Brit Mus (Nat Hist) Botany Dept*
Catalogues of bird drawings by WW Ellis and J Webber, of specimens from Cook's third voyage, and of animal drawings in the library of Sir J Banks, with some correspondence. *Ibid, Zoology Dept*

Annotated copy of his *Catalogus bibliothecae historico-naturalis Josephi Banks* (1796-1800) and related papers. *Ibid, General Libr*
Letters (102) to Sir J Banks 1782-1802. *Fitzwilliam Mus, Cambridge* (SG Perceval Colln)

[171] **DUNCAN, Andrew** (1744-1828)
Physician

Medical papers, incl lectures (3 sets) 1780-1800, and clinical observations (98 vols). *R Coll of Physicians, Edinburgh*

[172] **DUNLOP, James** (1795-1848)
Astronomer

Astronomical observations (4 vols). *Royal Soc*
Astronomical observations and memoranda made at Parramatta, Australia, 1832-40. *Mitchell Libr, Sydney*

[173] **DUTTON, Joseph Everett** (1877-1905)
Biologist

Papers incl diaries, field notes and records of experiments made during his expeditions to W Africa 1901-05. *Wellcome Hist Medical Libr* (MSS 2248-68)
Miscellaneous papers (1 folder), incl letters 1904, temperature charts and notes made during his last illness. *Liverpool School of Tropical Medicine*

DYER, see Thiselton-Dyer

[174] **DYSON, Sir Frank Watson,** KBE (1868-1939)
FRS, astronomer

Correspondence, observations and calculations, and administrative records as astronomer royal for Scotland 1906-10. *R Observatory, Edinburgh*
Correspondence and papers as astronomer royal 1910-33, incl papers rel to the administration of the Royal Observatory, Greenwich, the continuation of its work during World War I, its instruments, and exchanges with other scientific institutions. *R Greenwich Observatory, Herstmonceux*
Letters (116) to officers of the Royal Astronomical Society 1894-1900, and 21 to its Joint Permanent Eclipse Committee 1901-12; correspondence (9 items) with JH Reynolds 1920-35. *R Astronomical Soc*
Correspondence with GE Hale 1901-31. *California Inst of Technology, Pasadena*

See also Turner HH

[175] **EDDINGTON, Sir Arthur Stanley,** OM (1882-1944)
FRS, astronomer

Papers (56), incl MSS of articles and addresses, and lecture notes on the theory of relativity, astrophysics, celestial mechanics, philosophy, and stellar motions. *Private*
MS of his *Nature of the physical world* (1928); journal 1905-14; juvenile paper on astronomy. *Trinity Coll, Cambridge* (MSS 0.11.22, 0.11.25, Add b 48)
Papers. *Inst of Astronomy, Cambridge*
Correspondence with Viscount Cherwell 1913-33. *Nuffield Coll, Oxford* [NRA 16447]
Correspondence (14 items) with H Weyl 1918-44. *Swiss Federal Inst of Technology, Zurich*
Correspondence (13 items) with A Einstein 1919-29. *Inst for Advanced Studies, Princeton Univ*
Letters (11) to Sir J Larmor. *Royal Soc*

[176] **EDMONDSTON, Thomas** (1825-1846)
Naturalist

Notes for his *Flora of Shetland* (1845) from 1837. *Brit Mus (Nat Hist) Botany Dept*
Correspondence (*c*43 items) 1840-46. *Shetland Libr, Lerwick* [NRA 13501]

See also MacGillivray W

[177] **EGERTON, Sir Alfred Charles Glyn** (1886-1959)
FRS, chemist

Papers (*c*20 boxes) 1908-52, incl working papers on pyrometry, explosives, and steam tables, and lecture notes; college administrative correspondence and papers (2 files) 1936-52. *Imperial Coll, London*
Correspondence and papers (17 boxes), incl diaries 1929-30, 1943-59, diplomas, and papers concerning Royal Society business and foreign visits. *Royal Soc*
Correspondence with Viscount Cherwell 1914-*c*1949. *Nuffield Coll, Oxford* [NRA 16447]

[178] **ELIOT, Sir Charles Norton Edgcumbe,** GCMG (1862-1931)
Biologist and diplomat

Notebooks (4) and papers (1 box file) on nudibranch mollusca. *Brit Mus (Nat Hist) Zoology Dept*
Correspondence (1 vol) while ambassador to Japan 1924-25. *Public Record Office* (FO 800/255)
Correspondence with RAB Chamberlain 1903-08 and Sir FH May *c*1918. *Rhodes House Libr, Oxford* (MSS Afr s 589, Ind Ocn s 176(3)) [NRA 11225]
Correspondence with Lord Hardinge 1918-19.

Cambridge Univ Libr (Hardinge Papers 39-40)
Letters (10) to Sir G Bantock. *Birmingham Univ Libr*

[179] **ELLIS, John** (1714-1776)
FRS, naturalist

Correspondence and papers, incl a chemistry notebook 1751, 5 natural history notebooks 1755-76, notes for his *Natural history of . . . zoophytes* (1786), drafts of letters, and *c*400 letters addressed to him 1744-76. *Linnean Soc* [NRA 9516]
Papers (27) submitted to the Royal Society 1752-76. *Royal Soc*
Letters (52 ff) to DC Solander 1760-63. *Brit Libr* (in Add MS 29533)

▷S Savage, *Catalogue of the manuscripts in the Library of the Linnean Society of London. IV. Calendar of the Ellis Manuscripts*, 1948

See also Fothergill, Hales, Solander

[180] **ELLIS, Robert Leslie** (1817-1859)
Man of science

Diaries 1827-41, and *c*20 mathematical and other notebooks. *Trinity Coll, Cambridge* (MSS Add a 67, 82, 218-22)
Letters (37 ff) to Sir WFP Napier 1840-59. *Bodleian Libr, Oxford* (MS Eng lett c 245)
Correspondence (30 items) with JD Forbes 1835-59. *St Andrews Univ Libr* [NRA 13132]
Letters (30) to Lord Kelvin 1845-50. *Cambridge Univ Libr* (in MS Add 7342)
Letters (24) to CB Marlay 1849-58. *Nottingham Univ Libr* [NRA 10280]

[181] **ELTRINGHAM, Harry** (1873-1941)
FRS, entomologist

Entomological notebooks and paintings. *Hope Libr, University Mus, Oxford*

[182] **EVANS, Caleb** (1831-1886)
Geologist

Notes and drawings (3 vols) on the geology of the London basin and the south of England. *Brit Mus (Nat Hist) Palaeontology Dept*
Horizontal sections of the geology from Croydon to Oxted, Surrey, 1868. *Inst of Geological Sciences*

[183] **EVANS, Revd Lewis** (1755-1827)
FRS, mathematician

Papers and correspondence, incl notes for lectures (10 vols), notes on physics, astronomy and mathematics, catalogue of books of the Mathematical School at Christ's Hospital, and a paper on observatories. *History of Science Mus, Oxford* (MSS Evans 1-32) [NRA 9532]

[184] **EVEREST, Sir George** (1790-1866)
FRS, military engineer and surveyor

Correspondence, maps and other papers rel to the Survey of India. *Nat Archives of India, New Delhi*
Technical records and reports rel to the Survey of India. *Office of the Geodetic Branch, Survey of India, Dehra Dun, India*
Tables rel to the Survey of India. *Royal Soc*
Correspondence with the Earl of Ellenborough about the Survey of India 1830. *Public Record Office* (PRO 30/12/20/10) [NRA 21870]
Papers 1804-61. *R Artillery Inst* (MD/1300)
Letters (17) to the Royal Geographical Society 1861-62. *R Geographical Soc*

▷RH Phillimore, *Historical records of the Survey of India, IV: 1840 to 1843: George Everest*, 1958

[185] **EVERSHED, John** (1864-1956)
FRS, astronomer

Sketch books (16) of solar chromosphere and prominences 1889-1901, miscellaneous papers and correspondence, incl 13 letters to officers of the Royal Astronomical Society 1894-1900 and 28 to its Joint Permanent Eclipse Committee 1902-11. *R Astronomical Soc*
Correspondence with GE Hale 1904-37. *California Inst of Technology, Pasadena*

[186] **EWING, Sir James Alfred,** KCB (1855-1935)
FRS, engineer

Preliminary report on sea trials of the *Turbinia* 1897. *Science Mus, Newcastle upon Tyne* [NRA 9195]
Correspondence (18 items) with Lord Kelvin 1883-1903. *Cambridge Univ Libr* (in MS Add 7342)
Correspondence with the Earl of Balfour 1927-28. *Private* [NRA 10026]
Letters to the secretary of the Dundee Society of Engineers 1884-87. *Private*
Unpublished lecture on naval intelligence and Admiralty Room 40. *Private*

[187] **FAIRBAIRN, Sir William,** 1st Bt (1789-1874)
FRS, engineer

Letters, incl 39 to TR Robinson 1844-74 and 19 to Sir GG Stokes 1859-74. *Cambridge Univ Libr* (in MS Add 7656)
Letters (31) to WC and WJ Unwin 1856-73. *Imperial Coll, London* [NRA 10964]
Letters (15) to C Babbage 1840-55. *Brit Libr* (in Add MSS 37191-92, 37194, 37196)

[188] **FALCONER, Hugh** (1808-1865)
FRS, palaeontologist and botanist

Correspondence and papers, incl drawings, sections and plans of caves in Glamorgan, fossil drawings and notes rel to Glamorgan, Sicily, Kashmir and Tibet, 39 letters to MP Edgeworth 1836-46, 30 to his niece Grace McCall (later Lady Prestwich) 1854-64, and 22 to J Evans 1862-63. *Moray District RO, Forres* [NRA 20690]
Botanical papers (2 vols); illustrated unpublished memoir 'Coal fossils of Burdwan' 1833; 40 letters to Sir WJ Hooker 1837-49. *R Botanic Gardens, Kew*
Description of some fossil elephant remains in the Sedgwick Museum. *Geology Dept, Cambridge*
Letters (79) to Sir C Lyell 1843-62. *Edinburgh Univ Libr*

See also Prestwich

[189] **FARADAY, Michael** (1791-1867)
FRS, natural philosopher

Correspondence 1813-65 and papers, incl a diary 1820-62, laboratory notebooks (11 vols) 1820-62, notes of and for lectures 1810-50, MS of his *Chemical manipulation* (1827) and an annotated copy of his *Experimental researches in electricity* (1839-55). *Royal Inst* [NRA 9522]
Correspondence (*c*300 items) and 5 notebooks 1809-46, incl a diary of a journey to France and Italy with Sir H Davy 1813-14, notes of a walking tour in Wales 1819, geological notes on the Isle of Wight 1824, chemistry lecture notes 1816-19 etc. *Inst of Electrical Engineers* (Blaikley Colln) [NRA 20573]
'Glass furnace note book, Royal Institution'; MSS of his papers in *Philosophical Transactions;* album of diplomas etc; 40 letters to Sir JFW Herschel 1825-61. *Royal Soc*
Annotated copy of WT Brande's *Manual of chemistry* (1819). *Wellcome Hist Medical Libr* (MSS 2332-34)
Miscellaneous letters (184) addressed to him. *Private* [NRA 8908]
Correspondence (155 items) with CF Schönbein 1836-62. *Basle Univ Libr*
Correspondence (94 items) with W Whewell 1831-60. *Trinity Coll, Cambridge* [NRA 8804]
Correspondence (*c*60 items) with B Abbott 1812-19. *Private* [NRA 5334]
Correspondence (47 items) with J Plückner 1847-63. *Nat Research Council of Canada, Ottawa*
Letters (39) to C Babbage 1825-64. *Brit Libr* (Add MSS 37183-37200 *passim*)
Correspondence (37 items) with Sir GB Airy 1838-62. *R Greenwich Observatory, Herstmonceux* (in MSS 938-51)
Correspondence (30 items) with C and A De La Rive 1818-61. *Bibl Publique et Universitaire, Geneva*
Correspondence (19 items) with JD Forbes 1832-59. *St Andrews Univ Libr* [NRA 13132]
Correspondence with A Quételet 1837-65. *Académie Royale des Sciences, des Lettres et des Beaux-Arts de Belgique, Brussels*
Letters (14) to JB Dumas 1840-65. *Académie des Sciences, Inst de France, Paris*
Miscellaneous letters (13) 1820-63. *Burndy Corporation Libr, Norwalk, Connecticut*
Miscellaneous letters 1829-46. *R Military Academy Central Libr, Camberley*

▷ *Selected correspondence of Michael Faraday*, ed LP Williams, 2 vols, 1971, pp 1036-38

See also Airy, Davy, De La Rue, Sabine, Stokes, Thomson W

[190] **FARREN, Sir William Scott** (1892-1970)
FRS, engineer

Technical papers (5 boxes) 1916-70. *Churchill Coll, Cambridge*

[191] **FERGUSSON, Sir William,** 1st Bt (1808-1877)
FRS, surgeon

Diaries 1845-77, notes and other papers. *R Coll of Surgeons, London*
Letters (15) to William Blackwood & Sons 1842-75. *Nat Libr of Scotland* (MSS 4061-4331 *passim*)

[192] **FERRANTI, Sebastian Pietro Innocenzo Adhemar Ziani de** (1864-1930)
FRS, engineer

Papers and correspondence 1872-1930, incl working papers, diaries, patents, drawings, and business records of Ferranti companies. *Private* [NRA 13215]
Correspondence (22 items) with Lord Kelvin 1882-1906. *Cambridge Univ Libr* (in MS Add 7342)

See also Thomson W

[193] **FIELD, Joshua** (?1787-1863)
FRS, civil engineer

Papers (13 vols) 1811-63. *Science Mus Libr* [NRA 9524]

[194] **FIELDING, Henry Borron** (1805-1851)
Botanist

Papers, incl descriptions, plates and drawings rel to his herbarium. *Bodleian Libr, Oxford* (MSS Sherard 389-98) [NRA 6305]

Letters (122) to Sir WJ Hooker 1834-51. *R Botanic Gardens, Kew*

See also Gardner

[195] **FISHER, Sir Ronald Aylmer** (1890-1962)
FRS, geneticist and statistician

Notebooks *c*1943-1957 and correspondence 1919-59. *Adelaide Univ Libr*
Records of experiments on snails, dogs and mice. *Genetics Dept, Cambridge*
Miscellaneous correspondence on genetics. *Medical Research Council*

[196] **FITTON, William Henry** (1780-1861)
FRS, geologist

Correspondence (123 items) with C Babbage 1826-59. *Brit Libr* (in Add MSS 37183-37200)
Correspondence (23 items) with Sir JFW Herschel 1827-55; letters (14) to Sir JW Lubbock 1830-43. *Royal Soc*
Letters (15) to Earl Spencer 1814-20. *Private*

[197] **FITZGERALD, George Francis** (1851-1901)
FRS, physicist

Letters (182) to Sir OJ Lodge 1883-1900. *University Coll London* (MS Add 89)
Correspondence (24 items) with Lord Kelvin 1885-1901. *Cambridge Univ Libr* (in MS Add 7342)
A few papers 1881-1901, incl his report on the teaching of electrical engineering 1899. *Trinity Coll, Dublin* (MSS 2388, 4233, 4572) [NRA 19217]

[198] **FLAMSTEED, Revd John** (1646-1719)
FRS, astronomer

Papers and correspondence, incl his observations as astronomer royal 1675-1719, tables, and lectures at Gresham College 1681-84. *R Greenwich Observatory, Herstmonceux* [NRA 22822]
Navigational calculations and tables (6 sheets) 1706. *York Minster Libr*
Tables of the sun and moon 1700. *R Observatory, Edinburgh*
'Account of the . . . doctrine and practice of navigation' 1697. *Magdalene Coll, Cambridge* (Pepys Library MS 2184)
Letters from him, incl 56 to H Oldenburg 1669-77, 124 to A Sharp 1702-19 and 70 to R Towneley. *Royal Soc*
Letters (*c*35) to J Collins 1670-77. *Private*
Correspondence with E Bernard 1678-84 and T Smith 1700-09. *Bodleian Libr, Oxford* (MSS Smith *passim*)
Correspondence with Sir I Newton 1680-99. *Cambridge Univ Libr* (in MSS Add 3979, 4006)
Letters to Sir J Moore. *History of Science Mus, Oxford* (in MS Gunther 16)
Letters to J Hevelius 1681. *Trinity Coll, Cambridge* (in MS R5)
Papers by or rel to him collected by J and D Gregory. *Edinburgh Univ Libr* (MS Dc 1.61)

See also Molyneux, Newton I

[199] **FLEMING, Sir Alexander** (1881-1955)
FRS, bacteriologist

Papers and correspondence (120 vols) 1901-55, incl 46 notebooks 1901-53, 25 journals 1935-54, lectures, speeches and drafts 1940-55, and correspondence 1922-55. *Brit Libr* (Add MSS 56106-56225)
Correspondence, a bacteriological monograph, papers rel to the inoculation department of St Mary's Hospital, London, and other research and administrative topics. *Medical Research Council*
Some papers. *Wright-Fleming Inst, St Mary's Hospital Medical School, London*
Some papers. *Group 21 Medical Libr, Hartshill Orthopaedic Hospital, Stoke-on-Trent*

[200] **FLEMING, Sir John Ambrose** (1849-1945)
FRS, electrical engineer

Papers and correspondence 1856-1943, incl laboratory notebooks and other records of his experiments 1884-1925, patent specifications 1880-1915, correspondence and papers rel to his association with the Marconi Company 1899-1931, notes of lectures attended and for his own lectures, and his recollections of JC Maxwell 1879. *University Coll London* (MS Add 122) [NRA 14721]
Reports (10) on papers submitted to the Royal Society 1877-99; *c*20 letters from him. *Royal Soc*
Electrical laboratory notes. *Inst of Electrical Engineers* [NRA 20573]
Letters (13) to Sir J Larmor. *St John's Coll, Cambridge* [NRA 22852]

[201] **FLOREY, Howard Walter,**
Baron Florey of Adelaide and Marston (1898-1968)
FRS, pathologist

Papers and correspondence, incl his laboratory notebooks (41 vols) 1924-68 and other research papers, reports on penicillin and related correspondence, diaries 1926-68, pathology lectures, broadcasts, talks, papers as President of the Royal Society and other administrative papers, material rel to his published works, and family papers. *Royal Soc* [NRA 17536]

Correspondence and papers mainly rel to research topics. *Medical Research Council*
Papers and correspondence rel to the foundation of the John Curtin School of Medical Research. *Australian Nat Univ Archives, Canberra*
Dunham lectures, Harvard 1965. *Harvard Univ Medical School, Boston, Mass*
Correspondence with Viscount Cherwell 1938-55. *Nuffield Coll, Oxford* [NRA 16447]

▷G Macfarlane, *Howard Florey: the making of a great scientist*, 1979

[202] **FLOWER, Sir William Henry,** KCB (1831-1899)
FRS, zoologist

Notebooks (24), lecture notes, papers and correspondence rel mainly to cetacea, and drawings. *Brit Mus (Nat Hist) Zoology Dept*
Paper on collecting birds' eggs, list of principal additions by purchase to the Hunterian Museum 1806-62 and his annotated copy of Sir R Owen's *Descriptive catalogue of the osteological series*, II (1853). *R Coll of Surgeons, London* [NRA 9521]
Letters (52) to CE Fagan 1889-99; and letters to ACLG Günther 1869-98. *Brit Mus (Nat Hist) General Libr*
Letters (11) to TH Huxley 1862-95. *Imperial Coll, London*
Letters to OC March. *Sterling Memorial Libr, Yale Univ, New Haven*
Letters to JT Gulick. *American Philosophical Soc, Philadelphia*

[203] **FORBES, David** (1828-1876)
FRS, geologist

Chemical compilations *c*1845. *John Rylands Univ Libr, Manchester* [NRA 9194]
Letters (5) to Sir A Geikie 1861-71. *Edinburgh Univ Libr* (in MS Gen 524/3)

[204] **FORBES, Edward** (1815-1854)
FRS, naturalist

Field notebooks (18) on natural history 1844-54. *Brit Mus (Nat Hist) General Libr*
Field notebook on geology 1852-53 and other papers 1845-54. *Inst of Geological Sciences*
Journal in Norway 1833, sketches (11), and letters to Sir W Jardine and others. *Edinburgh Univ Libr* (in MSS Dk 6.20-21, MS Dc 6.91, MS Gen 524/3)
Papers rel to his lectures on fossil marine animals 1845 and the geographical distribution of organized beings 1851; correspondence (14 items) 1845-54. *Royal Inst*
Paper on the geology and flora of the Isle of Man *c*1836. *R Botanic Garden, Edinburgh*
Notes on the distribution of molluscs; correspondence (26 items) with Sir C Lyell and others 1839-54. *American Philosophical Soc, Philadelphia*
Papers read to the Linnean Society 1844, 1848; correspondence (6 items) 1836-52. *Linnean Soc*
Papers submitted to the Royal Society 1854, and 7 referee reports on the papers of others 1847-52. *Royal Soc*
Notes on British plants observed in Lycia, Turkey 1841-42. *Brit Mus (Nat Hist) Botany Dept*
'Note on the age of the Manx elk' 1843; correspondence (14 items) 1841-54. *Nat Libr of Scotland*
Letters (140) to Sir AC Ramsay 1844-54. *Imperial Coll, London*
Letters (31) to Sir JD Hooker and others 1843-54. *R Botanic Gardens, Kew*
Letters (25) to W Thompson 1840-51. *Manx Mus Libr, Douglas, Isle of Man* (MSS 2148A, 6968)
Correspondence (21 items) with TH Huxley 1846-54. *Imperial Coll, London*
Letters (11) to Sir WC and Lady Trevelyan *c*1842-54. *Newcastle Univ Libr* [NRA 12238]

▷PF Rehbock, 'Edward Forbes (1815-1854): an annotated list of published and unpublished writings', *J Soc Bibliog Nat Hist*, ix, pt 12, 1979, pp171-218

[205] **FORBES, George** (1849-1936)
FRS, engineer and astronomer

Correspondence and papers, incl an introductory survey of philosophy, diary 1882 and parts of journals, notes for lectures on astronomy 1923-29, correspondence (495 items), copies of correspondence (1 vol) 1871-74, and sketch books. *St Andrews Univ Libr* [NRA 13132]
Letters from him, incl 15 to officers of the Royal Astronomical Society 1873-1900 and 5 to its Joint Permanent Eclipse Committee 1909-10. *R Astronomical Soc*
Letters (17) to Sir GG Stokes 1871-82. *Cambridge Univ Libr* (in MS Add 7656)

See also Airy, Playfair, Smyth, Stewart

[206] **FORBES, James David** (1809-1868)
FRS, natural philosopher

Correspondence (67 vols) and papers 1815-68, incl journals, letter books, scientific diaries, commonplace books, lecture notes, records of experiments and MSS of published works. *St Andrews Univ Libr* [NRA 13132]
Correspondence (269 items) with Sir GB Airy 1836-67. *R Greenwich Observatory, Herstmonceux* (in MSS 938-51)
Correspondence (93 items) with Lord Kelvin 1845-61. *Cambridge Univ Libr* (in MS Add 7342)

Correspondence (69 items) with Sir JFW Herschel 1831-59. *Royal Soc*
Letters (12) to Sir GG Stokes 1850-56. *Cambridge Univ Libr* (in MS Add 7656)
Letters (11) to C Babbage 1831-55. *Brit Libr* (in Add MSS 37186-87, 37196, 37200)

See also Airy, Babbage, Baily, Brewster, Brisbane, Ellis RL, Faraday, Harcourt WVV, Harris, Herschel JFW, Jameson, Lloyd H, Lyell, Murchison, Phillips J, Playfair, Powell B, Sabine, Smyth, Stephenson R, Stewart, Tait, Talbot, Thomson W, Whewell

[207] **FORDYCE, George** (1736-1802)
FRS, physician

Papers on minerals and their arrangement, and notes on the chemical analysis of minerals in W Hunter's collection. *Hunterian Libr, Glasgow Univ*
Papers (10) submitted to the Royal Society. *Royal Soc*
Dissertation on inflammation *c*1770; student notes of his lectures. *R Coll of Physicians, Edinburgh*
Lectures 1785-87. *R Coll of Physicians, London*
Clinical lectures 1785. *Somerset RO, Taunton* (DD/PO/115)
Letters to Sir A Grant 1767-69. *Scottish Record Office* (GD 345/1171)
Letters to D Garrick. *Victoria & Albert Mus*

[208] **FORSTER, Edward** (1765-1849)
FRS, botanist

Notebooks (7), containing his botanical observations 1784-1848 and comments on botanical publications. *Brit Mus (Nat Hist) Botany Dept*
His annotated copies of botanical publications. *Passmore Edwards Mus, London*
Letters (26) to Sir JE Smith 1802-27. *Linnean Soc*
Letters (11) to Sir WJ Hooker 1831-46. *R Botanic Gardens, Kew*

[209] **FORSYTH, Andrew Russell** (1858-1942)
FRS, mathematician

Papers on pure and applied mathematics (4 boxes); college administrative correspondence and papers (2 files) 1913-31. *Imperial Coll, London*
Geometrical papers. *Royal Soc*
Letters (13) to K Pearson 1894-1936. *University Coll London*
Miscellaneous correspondence (27 items) 1892-1908. *Cambridge Univ Libr* (in MS Add 7656)

[210] **FOSTER, Henry** (1796-1831)
FRS, navigator

Records (2 vols) of pendulum experiments 1828-31. *R Astronomical Soc*
Magnetic experiments; magnetic and meteorological observations made on board HMS *Chanticleer* 1828-30. *Royal Soc*
Letters to Sir WE Parry. *Scott Polar Research Inst, Cambridge* (MS 438)

[211] **FOSTER, Sir Michael,** KCB (1836-1907)
FRS, physiologist

Notes and drawings (13 vols) on irises 1882-94. *Linnean Soc*
Letters (211) to TH Huxley and his wife 1865-1902. *Imperial Coll, London*
Letters (32) to Sir GG Stokes 1876-1900. *Cambridge Univ Libr* (in MS Add 7656)
Letters (13) to O Browning *c*1883-1904. *Hastings Public Libr*
Letters (11) to AB Kempe 1889-1902. *West Sussex RO, Chichester* [NRA 17595]
Miscellaneous letters (50). *Royal Soc*

See also Huxley TH

[212] **FOSTER, Samuel** (*d*1652)
Mathematician

Treatises on the sector, the cross-staff, J Napier's propositions for the solution of spherical triangles, and the geometrical square, 1626-29. *Bodleian Libr, Oxford* (MS Savile 102)

[213] **FOTHERGILL, John** (1712-1780)
FRS, physician and naturalist

Correspondence (86 items) and a few papers 1735-80. *Soc of Friends Libr, London* [NRA 20017]
Papers (7) submitted to the Royal Society. *Royal Soc*
Note on Siberian flora 1748. *Wellcome Hist Medical Libr* (in MS 2391)
Correspondence (40 ff) with EM da Costa 1746-79. *Brit Libr* (in Add MS 28537)
Letters (15) to C Alston 1737-50. *Edinburgh Univ Libr* (in MS La III 375)
Letters (13) to J Ellis *c*1774-76. *Linnean Soc* [NRA 9516]

▷BC and CC Corner, *Chain of friendship*, 1971

See also Baker JG

[214] **FOWLER, Alfred** (1868-1940)
FRS, astrophysicist

Correspondence 1906-35 and papers, incl his sun observations (5 vols) 1903-10, transit observations (1 vol) 1905-07, laboratory notebook rel to spectroscopic experiments 1906-13, spectrum analysis charts (3 vols), and 9 blueprints rel to the construction of an equatorial telescope, solar camera etc with related correspondence 1906-10. *Imperial Coll, London* [NRA 24526]

Letters from him, incl 52 to officers of the Royal Astronomical Society 1888-1900 and 14 to its Joint Permanent Eclipse Committee 1903-12. *R Astronomical Soc*

Correspondence (31 items) with N Bohr 1913-26. *Niels Bohr Inst, Copenhagen*

Correspondence with Viscount Cherwell 1919-33. *Nuffield Coll, Oxford* [NRA 16447]

Miscellaneous letters (17) 1904-25. *Royal Soc*

[215] **FOX, Sir John Jacob** (1874-1944)
FRS, chemist

Laboratory notebooks, correspondence and papers 1903-44. *Public Record Office* (DSIR 26)

Notebooks rel to his chemical research 1904-29, and correspondence (*c*40 items). *Private* [NRA 16920]

[216] **FOX, Robert Were** (1789-1877)
FRS, man of science

Letters (*c*13) to Sir E Sabine and others, with 66 addressed to him. *Royal Soc* [NRA 15294]

[217] **FRANKLAND, Sir Edward,** KCB (1825-1899)
FRS, chemist

Correspondence and papers *c*1840-1899, incl laboratory notebooks, reports and lecture notes. *Private*

Lecture notes on organic chemistry; transcript of his diary 1848-49. *Royal Soc*

Correspondence (26 items) with TH Huxley 1857-95. *Imperial Coll, London*

Correspondence (25 items) with Sir GG Stokes 1857-95. *Cambridge Univ Libr* (in MS Add 7656)

Letters (*c*20) to H Kolbe and others. *Deutsches Mus, Munich*

Letters to Sir JN Lockyer. *Exeter Univ Libr* (Norman Lockyer Observatory Colln)

Miscellaneous letters (24). *Royal Inst*

Miscellaneous letters (14). *Liverpool Univ Libr*

[218] **FREWEN, Thomas** (1704-1791)
Physician

'Encheiridion technicum in re medica' *c*1770. *Cambridge Univ Libr* (MS Add 6857)

Papers (2) submitted to the Royal Society. *Royal Soc*

[219] **GADDUM, Sir John Henry** (1900-1965)
FRS, pharmacologist

Correspondence and papers *c*1920-1965, incl 16 laboratory notebooks, diaries, working papers, lecture scripts and notes. *Royal Soc* [NRA 13899]

Personal diaries (2 vols) 1922-23. *Private*

Correspondence and papers mainly rel to research topics, committees and studentships. *Medical Research Council*

[220] **GALTON, Sir Francis** (1822-1911)
FRS, eugenist

Correspondence 1826-1911 and papers, incl diaries, travel notebooks and sketches 1830-1906, papers rel to his work on meteorology, heredity, statistics, psychometry, photography, fingerprints etc 1850-1910, and family history. *University Coll London* [NRA 19968]

Astronomical observations made in SW Africa 1850-51; letters, incl 230 to the Royal Geographical Society 1855-1902; correspondence with EJ Southon 1880. *R Geographical Society*

Papers rel to his anthropometric laboratory in South Kensington, incl instructions for making measurements, notes on craniometry, eye testing charts, fingerprint records and other case notes. *History of Science Mus, Oxford* (MSS Gunther 31-33) [NRA 9532]

Reports on papers submitted to the Royal Society and related correspondence (30 items). *Royal Soc*

Letters (49) to Macmillan & Co 1863-1909. *Brit Libr* (in Add MS 55218)

Letters to CR Darwin 1859-82. *Cambridge Univ Libr*

Letters (12) to Sir GG Stokes 1868-73. *Ibid* (in MS Add 7656)

Letters (11) to Lord Rayleigh 1876-1908. *Private*

Letters (10) to O Browning 1884-86. *Hastings Public Libr*

▷ M Merrington and J Golden, *A list of the papers and correspondence of Sir Francis Galton (1822-1911) held in the Manuscripts Room, The Library, University College London*, 1976

See also Bateson, Darwin CR, Darwin F, Darwin GH, Pearson, Poulton, Tyndall, Wallace

[221] **GARDNER, George** (1812-1849)
Botanist

Catalogue of Brazilian plants, drawings of Ceylon fungi; letters (130) to Sir WJ Hooker 1836-49. *R Botanic Gardens, Kew*

'Nova genera plantarum': plates and descriptions of plants, with HB Fielding.

Bodleian Libr, Oxford (MS Sherard 396) [NRA 6305]

See also Miers J

[222] **GASSIOT, John Peter** (1797-1877)
FRS, man of science

Record of early laboratory experiments; 13 letters from him. *Royal Inst*

Letters (18) to Sir JFW Herschel 1839-71; letters (10) to Sir JW Lubbock 1837-55. *Royal Soc*

Correspondence (16 items) with Sir GG Stokes 1861-*c*1869. *Cambridge Univ Libr* (in MS Add 7656)

[223] **GEIKIE, Sir Archibald,** OM, KCB (1835-1924)
FRS, geologist

Correspondence and papers 1851-1921, incl over 2,000 letters, geological and other notebooks, notes for lectures 1879-86, and papers and correspondence rel to the Royal Society's Sorby Research Fund 1909-14. *Edinburgh Univ Libr* (in MSS Gen 521-27, 694; MS Lyell 1)

Letters addressed to him as director general of the Geological Survey 1882-1901; letters from him 1865-1923, incl 223 letters to BN Peach 1867-1923, and 30 to FW Rudler, and an annotated map of Raasay *c*1860. *Inst of Geological Sciences* [NRA 18675]

Field notebooks. *Haslemere Educational Mus*

Notebook on the geology of Burdiehouse, Edinburgh 1853. *R Scottish Mus, Edinburgh*

Correspondence, incl *c*30 letters to him from GJP Scrope 1860-75, *c*45 from him to Sir RI Murchison and letters rel to his visit to N America 1897. *Geological Soc of London*

Correspondence with John Bartholomew & Son Ltd *c*1892-1920. *Private* [NRA 10823]

Letters to Macmillan & Co 1872-1924. *Brit Libr* (Add MS 55212)

Letters to the Revd G Gordon 1864-91. *Private* [NRA 22745]

Letters (17) from him and his son Roderick to O Browning *c*1893-1911. *Hastings Public Libr*

Letters (12) to Sir J and Lady Prestwich 1887-98. *Moray District RO, Forres* [NRA 20690]

Miscellaneous letters (80). *Royal Soc*

See also Darwin F, Davidson, Forbes D

[224] **GEIKIE, James** (1839-1915)
FRS, geologist

Correspondence (408 items) 1865-1914, incl 74 letters to BN Peach 1865-78; annotations of Sir AC Ramsay's topographical maps of Gibraltar 1876. *Inst of Geological Sciences* [NRA 18675]

Notes for his geology and Munro lectures 1913, and student notes (2 vols) of his geology lectures 1882-83. *Edinburgh Univ Libr* (MSS Dc 10.22-23)

Letters to his publishers 1876-92. *R Geographical Soc*

[225] **GIBSON, George Stacey** (1818-1883)
Botanist

Papers and correspondence rel to his *Flora of Essex* (1862), incl an early draft, proofs and *c*50 letters addressed to him. *Saffron Walden Mus*

Papers and correspondence, mainly genealogical. *Essex RO, Chelmsford* (D/Gi) [NRA 7968]

Miscellaneous correspondence, incl 6 letters from E Newman 1871-73. *Soc of Friends Libr, London*

[226] **GILBERT** (formerly **GIDDY**), **Davies** (1767-1839)
FRS, man of science

Correspondence and papers 1780-1839, incl engineering and mathematical notes, papers on road improvement schemes, almanacs containing memoranda, appointments, accounts etc 1780-1839, scrapbooks, and 148 letters from T Beddoes 1791-1808. *Cornwall RO, Truro* [NRA 9190]

Letters (32) addressed to him 1800-33. *Royal Inst of Cornwall, Truro* (Autograph Colln) [NRA 9188]

Correspondence (55ff) with Sir R Peel 1822-35. *Brit Libr* (in Add MSS 40347, 40394, 40397, 40411)

Correspondence (32 items) with Sir JFW Herschel 1826-30; letters (18) to Sir JW Lubbock 1830-35. *Royal Soc*

Letters (26) to C Babbage 1820-33. *Brit Libr* (in Add MSS 37182-85, 37205)

Letters (14) to J Hawkins 1793-1835. *West Sussex RO, Chichester* [NRA 10617]

[227] **GILL, Sir David,** KCB (1843-1914)
FRS, astronomer

Papers and correspondence, incl his heliometer observations (8 vols) 1877-82, account of a determination of the solar parallax (2 vols) 1877, a paper submitted to the Royal Astronomical Society 1879, 15 letters to its officers 1868-1900, 59 letters to AW Roberts 1893-1913 and 23 to GE Hale 1908-13. *R Astronomical Soc*

Papers and correspondence 1872-76 incl notes, calculations and determinations made at Dunecht Observatory and time determinations and observations made on Mauritius; notes on observations of Victoria and Sappho 1881; letters to Lord McLaren 1884-85. *R Observatory, Edinburgh*

Correspondence and papers as astronomer royal, Cape of Good Hope (1879-1907), incl administrative papers, letter books, records of the Anglo-German Boundary Survey 1896-1906 and the Geodetic Survey of South Africa 1899-1906. *R Greenwich Observatory, Herstmonceux*
Correspondence, incl *c*1,000 letters addressed to him *c*1870-1913. *R Geographical Soc*
Correspondence (21 items) with Lord Kelvin 1884-1907. *Cambridge Univ Libr* (in MS Add 7342)
Correspondence (16 items) with Sir GG Stokes 1886-89. *Cambridge Univ Libr* (in MS Add 7656)

See also Airy, Copeland, Darwin GH, Dewar, Larmor, Lockyer, Smyth, Turner HH

[228] **GLAISHER, James** (1809-1903)
FRS, astronomer and meteorologist

Papers and correspondence rel to balloon ascents 1862-70, a paper submitted to the Royal Astronomical Society 1862, 24 letters to its officers 1868-1900 and 44 to J Lee 1860-64. *R Astronomical Soc*

[229] **GLAZEBROOK, Sir Richard Tetley,** KCVO (1854-1935)
FRS, physicist

Papers and correspondence. *Public Record Office* (DSIR 10/20)
Letters to Macmillan & Co 1919-35. *Brit Libr* (Add MSS 55215-17)
Correspondence (27 items) with Lord Kelvin 1890-1905. *Cambridge Univ Libr* (in MS Add 7342)
Letters (23) to Sir OJ Lodge 1891-1915. *University Coll London* [NRA 20647]
Correspondence (17 items) with Lord Rutherford 1912-35. *Cambridge Univ Libr* (in MS Add 7653)
Correspondence on college affairs 1907-23. *Imperial Coll, London*

[230] **GLISSON, Francis** (1597-1677)
FRS, physician

Correspondence and papers, incl his lectures at the College of Physicians, medical and anatomical collections and observations, and letters addressed to him 1647-76. *Brit Libr* (in Sloane MSS 574B, 681, 1116, 2251, 2326, 3258, 3306-15)
Miscellaneous letters and papers. *R Coll of Physicians, London*

[231] **GODDARD, Jonathan** (1617-1675)
FRS, physician

'Collectanea de historia naturali' (1 vol), and chemical formulae (23ff). *Brit Libr* (in Sloane MSS 1139, 1159)
Papers (7) submitted to the Royal Society. *Royal Soc*

[232] **GODWIN-AUSTEN, Henry Haversham** (1834-1923)
FRS, naturalist

Survey of the glacier system of the Mustakh country 1864, notes on the Pangong Lake district of Ladakh 1867, journal of a tour in Assam 1872-73, and 30 letters to the Royal Geographical Society 1859-1918. *R Geographical Soc*
Annotated copy of his *Land and freshwater mollusca of India* (1899) with correspondence and drawings; notes and drawings of mollusca. *Brit Mus (Nat Hist) Zoology Dept*
Letters (14) to Sir C Lyell 1855-73. *Edinburgh Univ Libr*

[233] **GOSSE, Philip Henry** (1810-1888)
FRS, zoologist

Papers and MSS, incl an entomological letter book 1879-82, his 'Zoology of Jamaica' 1846, 'Fairy: a recollection' 1877, an autobiography, and a bibliography of his works down to 1869. *Cambridge Univ Libr* (MSS Add 7016-17, 7040, 7313, 7325 etc)
Notebooks on protozoa and sea anemones 1849, with a prospectus for his *Actinologia Britannica* (1858-60) and related papers. *Brit Mus (Nat Hist) Zoology Dept*
Notes, observations and tables for *The Aquarium* (1854); correspondence with the Revd C Kingsley 1853-58. *Brotherton Libr, Leeds Univ*
Notebook on the culture of plants, with water colours of specimens flowered by him 1863-75. *Brit Mus (Nat Hist) Botany Dept*
Notebook on the genital prehensors in butterflies 1882. *Ibid, Entomology Dept*
Sketches and drawings (200) by him and his correspondents of British sea anemones and corals 1839-61. *Horniman Mus, London*
Miscellaneous correspondence (45 items) *c*1856-60. *Edinburgh Univ Libr* (MS La II 425/22)

▷D Wertheimer, 'Philip Henry Gosse: science and revelation in the crucible', Toronto Univ PhD thesis 1977, pp549-54

[234] **GOULD, John** (1804-1881)
FRS, ornithologist

Notes for his *Birds of Europe* (1832-37); notes, correspondence and drawings (1 box) rel to Australian mammals and birds *c*1848-69. *Brit Mus (Nat Hist) Zoology Dept*
Papers and correspondence (2 vols) 1836-61 and drawings (8 vols) of Australian birds and mammals. *Mitchell Libr, Sydney* (ML MSS 547)
Papers, correspondence and drawings 1838-81, incl MS of his *Handbook to the*

birds of Australia (1865). *Nat Libr of Australia, Canberra* (in MSS 454, 587, 1519)
Notes and drawings (500ff) for his *Birds of Australia* (1840-48), 4 lists of Australian birds, and 9 letters to Sir W Jardine and A Newton 1835-61. *Newton and Balfour Libr, Zoology Dept, Cambridge*
Descriptions (2 vols) of 674 birds *c*1840, an unpublished paper on partridges 1844, and 22 drawings mainly of Australian birds 1831-66. *Blacker-Wood Libr, McGill Univ, Montreal*
Drawings (237) for his *Birds of Australia* (1840-48), and *c*1,300 others, with other artists, for his later publications. *Kenneth Spencer Research Libr, Univ of Kansas, Lawrence* (Jardine and Ellis Collns)
Drawings (204) of Australian, Asian and American birds and animals. *Private*
Drawings (56), correspondence with FT Buckland, Sir R Owen, HE Strickland and other naturalists, and 80 letters to his daughter. *Private*
Drawings of birds, and some letters from him 1857-58. *Sterling Memorial Libr, Yale Univ, New Haven*
Miscellaneous papers and correspondence (*c*36 items) 1857-77, incl accounts, receipts, 10 letters to AH Tulk and 5 to Sir R Barry. *State Libr of Victoria, Melbourne*
Drawings (30) and some correspondence. *Private*
Letters (43ff) to Sir W Jardine 1830-48. *R Scottish Mus, Edinburgh*
Letters (26) to Sir F McCoy. *Nat Mus of Victoria, Melbourne*
Letters (12) to FG Waterhouse. *South Australian Mus, Adelaide*

▷A McEvery, 'Collections of John Gould manuscripts and drawings', *La Trobe Library Journal*, i, pt 2, Oct 1968, pp17-31; *idem*, 'John Gould's contribution to British art', *Art Monograph 2*, Australian Academy of the Humanities, 1973

See also Hodgson, Jardine

[235] **GRAHAM, Thomas** (1805-1869)
FRS, chemist

Notebooks (24) 1835-64, essays, lecture notes and other papers 1821-68. *Wellcome Hist Medical Libr* (MSS 2551-84)
Laboratory notebooks 1832-35. *Andersonian Libr, Strathclyde Univ, Glasgow*
Letters from him (57) on college affairs 1837-56. *University Coll London*
Letters (19) to Lord Playfair 1839-67. *Imperial Coll, London* [NRA 11556]

[236] **GRAY, John Edward** (1800-1875)
FRS, naturalist

Papers and correspondence *c*1830-1874, incl an autobiographical journal *c*1862 and drafts of an autobiography *c*1862-63, notes (3 vols) on rodent specimens in the Leiden Museum, and catalogues, lists, descriptions and notes rel to zoological specimens in the British Museum. *Brit Mus (Nat Hist) Zoology Dept*
Notes (1 vol) for his lectures on plants and insects *c*1819-20; correspondence with ACLG Günther. *Ibid, General Libr*
Diplomas 1823-74; correspondence, incl letters (2 vols) to RM Crawshay 1862-73. *Brit Libr* (Add MSS 29533, 29717, 40140-41)
Correspondence (487 items) 1824-75, with some miscellaneous papers. *American Philosophical Soc, Philadelphia* [NRA 9728]
Letters (47) to Sir WJ Hooker 1826-65. *R Botanic Gardens, Kew*
Letters (28) to Margaret Gatty 1863-69. *Sheffield Central Libr* (MD 2134) [NRA 15568]
Letters (12) to W Swainson 1828-31. *Linnean Soc*
Letters (11ff) to Sir W Jardine 1834-47. *R Scottish Mus, Edinburgh*
Letters (*c*10) to Sir WC Trevelyan *c*1848-1871. *Newcastle Univ Libr* [NRA 12238]

▷AE Gunther, 'A note on the autobiographical manuscripts of John Edward Gray (1800-1875)', *J Soc Bibliog Nat Hist*, vii, pt 1, 1974, pp35-76

See also Hardwicke, Hodgson

[237] **GREAVES, William Michael Herbert** (1897-1955)
FRS, astronomer

Correspondence and papers, incl papers rel to transit circles, wireless time, clocks, instruments, spectroscopy, photometry, and photography, and administrative records as astronomer royal for Scotland 1938-55. *R Observatory, Edinburgh*

[238] **GREEN, Arthur George** (1864-1941)
FRS, chemist

Notebooks, papers and letters 1895-1941. *Brotherton Libr, Leeds Univ*

[239] **GREENOUGH, George Bellas** (1778-1855)
FRS, geographer and geologist

Papers and correspondence *c*1790-1855, incl journals, commonplace books, notebooks on architecture, cartography, ethnology, geography, geology; papers rel to his house in Regent's Park; Egyptological correspondence and papers; letters (16) to the Society for the Diffusion of Useful Knowledge 1833-44. *University Coll London*
Papers and correspondence, incl his lectures on stratification *c*1813, notes on mineralogical nomenclature, description, with R Phillips

and L Horner, of fuller's earth pits at Nutfield, papers rel to Geological Society business 1807-54, catalogue of the MSS etc in his library, and annotated copies of maps and other publications. *Geological Soc of London*

Papers and correspondence, particularly rel to the Geological Society of London. *Cambridge Univ Libr* (MS Add 7918)

Lists and descriptions of rocks, minerals and fossils in the museum of the Geological Society of London. *Inst of Geological Sciences*

'Report on a proposal of Saxe Bannister on maps for the new Houses of Parliament' 1849. *R Geographical Soc*

Catalogue of his collections (6 vols), with indexes and classification schemes (3 vols). *Geology Dept, University Coll London*

▷ J Golden, *A list of the papers and correspondence of George Bellas Greenough (1778-1855) held in the Manuscripts Room, University College London Library*, 1981

See also Horner, Jameson

[240] **GREGORY, David** (1661-1708)
FRS, astronomer and mathematician

Treatises and notes on mathematics and astronomy, lecture notes, and student notes of his lectures. *Edinburgh Univ Libr* (in MSS Dc 1-7, MSS La III 170, 175, 570)

Treatises on mathematics; student notes (3 sets) of his lectures. *St Andrews Univ Libr*

Notes on Sir I Newton's *Principia*, and family papers and correspondence (63 items). *Royal Soc*

Notes on Sir I Newton's *Principia*, 'Institutiones astronomicae', and mathematical tables. *Christ Church, Oxford* (MSS 131, 133)

Letters (10) to Sir I Newton 1684-1702. *Cambridge Univ Libr* (MS Add 3980)

See also Flamsteed

[241] **GREGORY, James** (1638-1675)
FRS, mathematician

[James Gregory's own papers, catalogued by his nephew David Gregory, are now dispersed. Many have been located and published in translation by HW Turnbull in *James Gregory: tercentenary memorial volume, containing his correspondence with John Collins, and his hitherto unpublished mathematical manuscripts*, 1939. This volume also incorporates the Gregory-Collins correspondence published by SP Rigaud in *Correspondence of scientific men of the seventeenth century*, 1841-62, from the collection of the Earl of Macclesfield.]

Mathematical papers (221 items). *Edinburgh Univ Libr* (MS Dc 1.61; MS Dk 1.22)

Mathematical papers (3). *Brit Libr* (in Sloane MS 3208)

Correspondence (24 items) with J Collins 1667-75. *Private*

Correspondence (57 items) with J Collins 1668-75. *St Andrews Univ Libr*

Miscellaneous papers and correspondence, incl some correspondence with J Collins 1669-74. *Royal Soc*

See also Flamsteed

[242] **GREGORY, James** (1753-1821)
Professor of medicine

Case notes (2 vols) 1780-81, clinical notes and lectures (2 vols) 1779-81, lectures on physiology (1 vol) *c*1775-80, and student notes (8 sets) of his lectures 1796-1817. *R Coll of Physicians, Edinburgh* [NRA 16015]

Case notes 1785-86, medical notebook *c*1795, essay on power, and student notes (15 sets) of his lectures 1777-1813. *Edinburgh Univ Libr* (MSS La III 788-89, MSS Gen 553D, 691-93D, and MSS Dc *passim*)

Case notes 1785. *R Coll of Physicians, London*

See also Hope TC

[243] **GREGORY, John** (1724-1773)
FRS, physician

Personal accounts 1769-73, and lectures on the institutions of medicine 1773, with 2 sets of student notes 1769-73. *Edinburgh Univ Libr* (MS La III 793; MSS Dc 6.125-26, 7.116)

Notes for his lectures on medicine *c*1766, with 7 sets of student notes *c*1766-1772. *R Coll of Physicians, Edinburgh* [NRA 16015]

Letters to Sir A Grant 1759. *Scottish Record Office* (GD 345/1180)

[244] **GREGORY, John Walter** (1864-1932)
FRS, geologist

Report on his visit to the Caribbean 1899. *Brit Mus (Nat Hist) General Libr*

Plans (21), sketch maps and notes rel to his travels in Africa *c*1912. *Glasgow Univ Archives* [NRA 18909]

Miscellaneous correspondence and papers, incl his hypsometric and astronomical observations in Yunnan and Tibet 1913. *R Geographical Soc*

[245] **GREGORY, William** (1803-1858)
Chemist

Notebooks (6) rel to his collection of diatom slides *c*1840-50. *Brit Mus (Nat Hist) Botany Dept*

[246] **GREVILLE, Robert Kaye** (1794-1866)
Botanist

Catalogue of diatoms, and correspondence with GAW Arnott and others. *Brit Mus (Nat Hist) Botany Dept*
Letters (59) to Sir WJ Hooker 1820-56. *R Botanic Gardens, Kew*
Correspondence (29 ff) with Sir W Jardine 1822-37. *R Scottish Mus, Edinburgh*
Letters (26) to Sir WC Trevelyan *c*1826-1863. *Newcastle Univ Libr* [NRA 12238]

See also Arnott

[247] **GREW, Nehemiah** (1641-1712)
FRS, plant physiologist

Papers and correspondence, incl his medical observations, catalogue of plants, catalogue of the Royal Society's museum at Gresham College, Egyptological collections, Greek and Latin glossary 1656, Latin translation of part of his *Anatomy of plants* (1682). *Brit Libr* (in Sloane MSS 1926-4076)
Papers (5) on botany and agriculture. *Royal Soc*
Prescriptions. *R Coll of Physicians, London*
Letters (10) to M Lister 1673-81. *Bodleian Libr, Oxford* (in MSS Lister 33-34)

[248] **GRIFFITH, William** (1810-1845)
Botanist

Papers and correspondence, incl Asian travel journals (3 vols) and itineraries (3 vols), botanical notes and index (12 vols), MSS of unpublished and posthumously published works; letters (12) to Sir WJ Hooker 1840-44. *R Botanic Gardens, Kew*
Report on the products of Afghanistan, journal of a visit to Assam, and 23 drawings of fishes from Afghanistan and India 1839-40. *India Office Libr and Records* (MSS Eur D 159, 517; NHD 7/1117-39)
Notes and drawings for articles published in the Linnean Society's *Transactions* (1836-47). *Linnean Soc*
Letters (16) to R Brown 1833-44 and (5) to RH Solly 1840-41. *Brit Libr* (in Add MS 32441)

[249] **GROVE, Sir William Robert** (1811-1896)
FRS, man of science

Papers and correspondence (over 800 items). *Royal Inst*

See also Lyell, Sabine, Talbot

[250] **GULL, Sir William Withey,** 1st Bt (1816-1890)
FRS, physician

Clinical notes on effusions of blood into the brain 1846-49, his interleaved copy of *Pharmacopoeia Guyensis* (1827) annotated with prescriptions 1838, notes taken by him at lectures 1840-41, and student notes on his lectures 1852-53. *Wellcome Hist Medical Libr* (MSS 2136, 2140, 2654-55, 3079-80)
Letters to Sir HW Acland and his daughter 1870-90. *Bodleian Libr, Oxford* [NRA 22893]

[251] **GÜNTHER, Albert Charles Lewis Gotthilf** (1830-1914)
FRS, zoologist

Correspondence (*c*55 boxes) 1850-1914 and papers, incl diaries and notebooks 1862-1903, papers as keeper of the Department of Zoology 1875-95, MSS of published works and corrected proofs. *Brit Mus (Nat Hist) General Libr*
Diary and correspondence (4 vols) 1847-95. *Brit Libr* (Add MSS 54488-91)
Papers and correspondence rel to the British Association, the Linnean Society and the Scientific Relief Fund 1896-1900. *History of Science Mus, Oxford* (MSS Gunther 11-13) [NRA 9532]
MS on central American reptiles and frogs, and guide to an elementary knowledge of British fishes. *Linnean Soc*
Correspondence (72 ff) 1861-72. *National Mus of Natural Hist, Leiden*
Letters (29) to the Eyton family 1869-1901. *Birmingham Univ Libr* (Eyton Letters 52-57, 109-31) [NRA 13199]
A few letters to CR Darwin, incl observations and queries rel to Darwin's *Descent of man* (1871). *Cambridge Univ Libr*

See also Couch, Darwin CR, Flower, Gray, Gunther RWT, Harmer, Hodgson, Hooker JD, Mivart, Murray, Newton A, Powys

[252] **GUNTHER, Robert William Theodore** (1869-1940)
Zoologist and historian of science

Correspondence and papers 1909-40, incl letter books and papers rel to his curatorship of the Museum of the History of Science, notes on herbals, lectures, notes and articles rel to his historical research, annotated copy of his *Early science in Oxford*, pts I-IV (1920-25). *History of Science Mus, Oxford* (in Gunther MSS *passim;* Old Ashmolean Letter Books) [NRA 9532]
Correspondence and papers mainly rel to his work on the history of science, including notes for a lecture 1897, MSS of his published works, and letters addressed to

him 1900-22. *Magdalen Coll, Oxford* (MSS 557, 564-66, 570-71, 580-91, 610)
Correspondence and papers incl notebooks, papers and drawings rel to his work on marine zoology in Britain, Italy and Persia 1885-1924, letters to him at Oxford 1886-92, Oxford notebook 1888-1920, notes on Anatolia and the Transcaucasus 1898, unpublished work on the zoology of S Italy *c*1908-09, notes for a biography of his father ACLG Günther and annotated copies of his father's *Catalogue of the fishes in . . . the British Museum* (1859-70). *Brit Mus (Nat Hist) General Libr*
Notes on the Lady Hippisley collection of minerals. *Ibid, Mineralogy Dept*
Notes on EA Minchin's lectures on sponges 1890. *Linnean Soc*

[253] **HADFIELD, Sir Robert Abbott,** Bt (1858-1940)
FRS, metallurgist
Correspondence and miscellaneous personal papers 1894-1934. *Sheffield Central Libr* (MD 4032-44) [NRA 22848]
Diaries 1875-78. *Private*
Bibliography of the effects of low temperatures on metals 1869-1933. *Royal Soc*
Correspondence with GE Hale 1916-36. *California Inst of Technology, Pasadena*

[254] **HALDANE, John Scott** (1860-1936)
FRS, physiologist
Scientific, personal and family papers and correspondence 1879-1936. *Nat Libr of Scotland*
Papers and correspondence 1878-1935, incl reports, papers and correspondence on gas warfare and mining, MSS of published and unpublished books, articles and lectures, and letters from his family. *Woodward Biomedical Libr, Univ of British Columbia, Vancouver* (Sinclair Colln) [NRA 14062]
Paper on heat engines. *Private* [NRA 10026]
Correspondence (13 items) with AV Hill 1929-30. *Churchill Coll, Cambridge*
Miscellaneous correspondence. *Medical Research Council*

[255] **HALES, Revd Stephen** (1677-1761)
FRS, physiologist and inventor
Papers rel to his work on ventilators. *Science Mus Libr*
MS of his *Philosophical experiments* (1739); papers (10) submitted to the Royal Society. *Royal Soc*
Letters (15) to J Ellis 1752-58. *Linnean Soc*
Abstracts of letters from him 1722-60. *Soc for Promoting Christian Knowledge, London*
▷DGC Allan and RE Schofield, *Stephen Hale, scientist and philanthropist,* 1980, pp178-90

[256] **HALL, Sir Arthur John** (1866-1951)
Physician
Notes and case papers, particularly rel to encephalitis and cramp. *Sheffield Univ Libr*
Correspondence and papers, mainly rel to encephalitis. *Medical Research Council*

[257] **HALL, Sir James,** 4th Bt (1761-1832)
Geologist and chemist
Diaries (13), correspondence, papers and drawings rel to chemical experiments 1774-1831. *Scottish Record Office* (GD 206/2, 237/37)
Diaries (9) 1783-91, and correspondence 1804-24. *Nat Libr of Scotland* (MSS 584, 3220-21, 3813, 4002, 6324-32)

[258] **HALLEY, Edmond** (1656-1742)
FRS, astronomer
Papers, incl observations (4 vols), computations and tables. *R Greenwich Observatory, Herstmonceux* (MSS 77-89) [NRA 22822]
Journals (3 vols) of his voyages in the *Paramore* in the Atlantic 1699-1700 and the English Channel 1701. *Brit Libr* (Add MS 30368)
Papers (61) submitted to the Royal Society, and miscellaneous correspondence (15 items) 1676-97. *Royal Soc*
'Compendium logicae et ethicae', lectures in Latin, and correspondence. *Private*
Papers on solar and lunar eclipses (28pp) *c*1692. *Private*
Tables rel to Saturn 1678, and a paper on tides 1687. *Brit Libr* (in Sloane MSS 1030, 1782)
'On the study of navigation' 1696. *Magdalene Coll, Cambridge* (Pepys Library MS 2185)
Letters (34) to J Burchett 1698-1702. *Public Record Office* (in Adm 1)
Letters (9) to Sir I Newton 1686-87. *King's Coll, Cambridge* (Keynes Newton MS 97)
Correspondence with Sir I Newton 1695-1725. *Cambridge Univ Libr* (MS Add 3982)
▷EF MacPike, *Correspondence and papers of Edmond Halley,* 1932

See also Stratford

[259] **HALLIBURTON, William Dobinson** (1860-1931)
FRS, physiologist
Notebooks (48) 1874-1902, incl notes on lectures attended 1875-82 and records of experiments 1883-1902. *Wellcome Hist Medical Libr* (MSS 2672-2719)

[260] **HAMEY, Baldwin** (1600-1676)
Physician
Papers and correspondence, incl the original MS of 'Bustorum aliquot reliquiae' 1628-76, 'Universa medicina' 1624, anatomy lectures

(91 ff) 1648, and medical notes. *R Coll of Physicians, London*

See also Harvey

[261] **HAMILTON, David James** (1849-1909)
FRS, pathologist

Correspondence (4 items) with Sir GG Stokes 1884-85. *Cambridge Univ Libr* (in MS Add 7656)
Student notes (2 vols) on his pathology lectures 1880-81. *Wellcome Hist Medical Libr* (MSS 2762-63)

HAMILTON, F, see Buchanan, FH

[262] **HAMILTON, Sir William Rowan** (1805-1865)
Mathematician

Correspondence (*c*2,300 items) and papers 1820-65, incl mathematical notebooks, astronomical observations, lecture notes, drafts of published works and poems. *Trinity Coll, Dublin* (MSS 1492-93, 7762-76) [NRA 20076]
Letters (22) to Lord Kelvin 1845-49. *Cambridge Univ Libr* (in MS Add 7342)
Letters (10) to Sir JW Lubbock 1831-38. *Royal Soc*

See also De Morgan, Herschel JFW, Lloyd H, Lubbock JW, Robinson TR, Tait

[263] **HAMMOND, Sir John** (1889-1964)
FRS, geneticist

Correspondence and papers, incl notebooks and typescripts of broadcasts. *Private*

[264] **HAMPSON, Sir George Francis,** 10th Bt (1860-1936)
Entomologist

Notebooks rel to insects collected in the Nilgiri Hills, India, *c*1888, and 5 unpublished volumes on the noctuidae intended to supplement his *Catalogue of the lepidoptera phalaenae in the British Museum. Brit Mus (Nat Hist) Entomology Dept*

[265] **HANBURY, Daniel** (1825-1875)
FRS, pharmacist

Letter books (14) rel to the writing of *Pharmacographia* (1874), and business records of Allen & Hanbury Ltd. *Private* [NRA 0384]
Correspondence and papers 1856-75, mainly rel to Indian, Chinese and S American materia medica. *Pharmaceutical Soc of Great Britain*
Letters (90) to Sir WJ Hooker 1848-65. *R Botanic Gardens, Kew*

See also Spruce

[266] **HARCOURT, Augustus George Vernon** (1834-1919)
FRS, chemist

Letters (33) to Sir GG Stokes 1866-96. *Cambridge Univ Libr* (in MS Add 7656)
Letters (23) to Sir J Conroy 1873-84. *Balliol Coll, Oxford* [NRA 22851]

[267] **HARCOURT, Revd William Venables Vernon** (1789-1871)
FRS, chemist

Notebook entitled 'Experiments on glass No 1' *c*1863-70. *Scientific Periodicals Libr, Cambridge*
Correspondence (*c*250 ff) with his family *c*1801-1869. *Bodleian Libr, Oxford* (in MSS Harcourt dep 601-02, 670) [NRA 3679]
Correspondence (145 items) with Sir GG Stokes *c*1862-1871. *Cambridge Univ Libr* (in MS Add 7656)
Correspondence (41 items) with JD Forbes 1831-66. *St Andrews Univ Libr* [NRA 13132]
Correspondence (25 items) with Sir R Peel 1841-43. *Brit Libr* (in Add MSS 40494, 40530-31)
Correspondence (12 items) with C Babbage 1831-39. *Ibid* (Add MSS 37186-37200 *passim*)

[268] **HARDWICKE, Thomas** (1755-1835)
FRS, naturalist

Correspondence and papers, incl journal of his voyage to India 1815, meteorological diaries 1813-33, zoological notes and descriptions, and drawings rel to Indian botany and zoology. *Brit Libr* (Add MSS 9868-9912, 10985-11032 *passim*, 46917)
Correspondence and accounts 1810-34, and papers rel to JE Gray's *Illustrations of Indian zoology* 1830-34. *Public Record Office* (C 103/192)
Drawings (32 vols) collected by him mainly rel to Asiatic zoology. *Brit Mus (Nat Hist) Zoology Dept*
Correspondence, and drawings and descriptions of plants in India 1796 and S Africa 1812. *Ibid, Botany Dept*
Notes on drawings by W Farquhar (6 vols) of the fauna and flora of Malacca. *Ibid, General Libr*
Drawings (1 vol), with Mrs Duncan Campbell, rel to Indian zoology and botany 1822. *Brit Mus (Nat Hist), Tring*
Drawings (96 in 1 vol) of Indian birds

1794-1803. *India Office Libr and Records* (NHD 39)
Papers and drawings rel to Indian animals *c*1813-23; letters (15) to Sir JE Smith 1804-19. *Linnean Soc*
Letters to the Marchioness of Hastings 1819-25. *Private* [NRA 15459]

▷ WR Dawson, 'On the history of Gray and Hardwicke's *Illustrations of Indian zoology*, and some biographical notes on General Hardwicke', *J Soc Bibliog Nat Hist*, ii, pt 3, 1946, pp55-69

See also Roxburgh, Wallich N

[269] **HARDY, Godfrey Harold** (1877-1947)
FRS, mathematician

Notebooks and papers. *Trinity Coll, Cambridge* (Add MSS b 55-56)
Asymptotic formulae, with S Ramanujan; papers on Waring's Problem, with JE Littlewood. *Cambridge Univ Libr* (MSS Add 6982-83)
Correspondence (32 items) with Bertrand Russell 1903-24. *McMaster Univ Libr, Hamilton, Canada* [NRA 12092]
Correspondence in connection with the British Association Mathematical Tables Committee 1931-36. *Bodleian Libr, Oxford* [NRA 21373]
Letters to EC Titchmarsh. *New Coll, Oxford* [NRA 20207]

[270] **HARMER, Sir Sidney Frederic,** KBE (1862-1950)
FRS, zoologist

Bibliography (8 vols) of the polyzoa, and drawings. *Brit Mus (Nat Hist) Zoology Dept*
Letters to ACLG Günther 1908-13; correspondence with the Countess of Denbigh *c*1912. *Ibid, General Libr*
Letters (59) to Macmillan & Co 1894-1928. *Brit Libr* (in Add MS 55223)
Letters (43) to O Browning 1882-1909. *Hastings Public Libr*
Correspondence (*c*40 items) with HE Forrest 1929-31. *Linnean Soc*
Correspondence with AGB Cherry-Garrard 1913-23. *Scott Polar Research Inst, Cambridge*

[271] **HARRIOT, Thomas** (1560-1621)
Mathematician and astronomer

Astronomical papers and drawings, papers on algebra, geometry, the calendar, conic sections, physics, chemistry, and miscellaneous calculations. *Private*
Miscellaneous mathematical papers, incl calculations and annotations, and correspondence 1610-19. *Brit Libr* (Harleian MS 6083; Add MSS 6782-89)
Copy of Vitellio's *Opticae libri decem*, ed F Risner (1572), containing MS tables of refraction written by Harriot. *Oslo Univ Libr*

▷ *Thomas Harriot, Renaissance scientist*, ed JW Shirley, 1974

[272] **HARRIS, Sir William Snow** (1791-1867)
FRS, electrician

Letters (31) to C Babbage 1844-56. *Brit Libr* (in Add MSS 37193-95, 37197)
Correspondence (15 items) with JD Forbes 1831-42. *St Andrews Univ Libr* [NRA 13132]
Correspondence *c*1837. *Private* [NRA 11184]

[273] **HARRISON, John** (1693-1776)
Horologist

Papers and drawings, incl his own descriptions of his clocks and chronometers 1730-71. *Guildhall Libr, London* (MSS 3972/1-2, 3973, 6026/1-6)
Contemporary copies of his memorial, testimonials etc, claiming reward for his chronometer 1736. *Lincolnshire Archives Office* (Monson Papers 7/19) [NRA 9511]

[274] **HARTLEY, Sir Harold,** GCVO (1878-1972)
FRS, engineer

Correspondence and papers (385 boxes), incl gas warfare papers 1915-40, mineral oil papers 1939-40, British Airways etc papers 1943-72, and correspondence 1890-1970. *Churchill Coll, Cambridge*
Early notebooks. *Physical Chemistry Laboratory, Oxford*

[275] **HARTLEY, Sir Walter Noel** (1847-1913)
FRS, chemist

Catalogue (4 vols) of spectra 1895-1907. *Science Mus Libr* [NRA 9524]
Notebooks (2) on the uric acid group 1894, 1904. *University Coll Dublin*
Letters (27) to Sir GG Stokes 1873-99. *Cambridge Univ Libr* (in MS Add 7656)

[276] **HARTLIB, Samuel** (*c*1600-1662)
Man of science

Correspondence and papers (72 vols) 1622-62, incl financial papers, treatises, and papers relating to husbandry, projects, mechanical inventions, astronomy, natural philosophy, education, foreign and colonial affairs, the Revd J Durie and the union of Protestant churches. *Sheffield Univ Libr* (50 H) [NRA 23591]

Correspondence (21 items) with J Pell 1655-59. *Brit Libr* (in Add MSS 4279-80, 4377, 4408, 4431)
Transcript of his letters (49) to J Worthington 1655-62 made by T Baker (originals lost). *Cambridge Univ Libr* (MM.1.40)

See also Oldenburg

[277] **HARVEY, William** (1578-1657)
Physician

'Prelectiones anatomiae' 1616; 'De musculis, motu locali &c' 1627. *Brit Libr* (Sloane MSS 230A, 486)
Prescriptions *c*1630-1656. *Ibid* (in MS Cotton Vitell C III; Sloane MSS 206A, 520; Add MS 36308)
Prescriptions and medical advice to J Aubrey. *Bodleian Libr, Oxford* (in MSS Aubrey A13, 21)
Correspondence (1 vol) rel to B Hamey, and 11 letters to Lord Feilding 1636. *R Coll of Physicians, London*

[278] **HARWOOD, Sir Busick** (1745-1814)
FRS, anatomist

Lectures attributed to him. *Botany School, Cambridge*

[279] **HATCHETT, Charles** (?1765-1847)
FRS, chemist

Correspondence (93 items) 1798-*c*1837 and papers, incl observations on the action of acids on animal substances 1809 and chemical notes. *University Coll of Wales, Swansea* (Royal Inst of South Wales MSS) [NRA 17990]
Papers (17) submitted to the Royal Society. *Royal Soc*
Letters (44) to the Revd T Rackett and family 1790-1838. *Dorset RO* (Solly Papers) [NRA 8506]

[280] **HAWORTH, Adrian Hardy**
(1767-1833)
Naturalist

MS of his *Lepidoptera Britannica*. *Hancock Mus, Newcastle upon Tyne*
Letters (64ff) to the Revd HT Ellacombe. *R Botanic Gardens, Kew*
Letters (28) to M van Marum 1814-25. *Netherlands Soc of Sciences, Haarlem*

[281] **HAWORTH, Sir Walter Norman**
(1883-1950)
FRS, chemist

Reports and correspondence rel to uranium compounds and early work on atomic energy. *Atomic Energy Research Establishment, Harwell*
Notebooks and papers on the published work of the department. *Chemistry Dept, Birmingham Univ*

[282] **HEBERDEN, William** (1710-1801)
FRS, physician

Papers, incl 3 Gulstonian lectures on poisons 1749, MS of his *Introduction to the study of physic* (1929), index to his commentaries on the history of diseases, index to materia medica, essays and notes. *R Coll of Physicians, London*
Papers (3) submitted to the Royal Society. *Royal Soc*

[283] **HENRY, William** (1774-1836)
FRS, chemist

Notes taken at animal economy lectures of J Allen 1795-96; notes for his chemistry lectures 1798-99 and 1804-05; notes 1811-29 mainly for successive editions of his *Elements of experimental chemistry*. *John Rylands Univ Libr, Manchester* [NRA 9194]
Papers (3) submitted to the Royal Society. *Royal Soc*
Letters (10) to C Babbage 1829-35. *Brit Libr* (Add MSS 37184-37200 *passim*)

[284] **HENSLOW, Revd John Stevens**
(1796-1861)
Botanist

Papers, incl his catalogue of the library of the Cambridge botanical garden, notes for lectures, annotated copy of R Relhan's *Flora Cantabrigiensis*, 3rd edn (1820), and letters addressed to him (1 vol). *Botany School, Cambridge* [NRA 9541]
Proof corrections to his study of the geology of Anglesey, drawings, sections and maps. *Christ's Coll, Cambridge* [NRA 9499]
Miscellaneous papers and correspondence, mainly rel to scientific societies, *c*1820-1852. *Suffolk RO, Ipswich* (HD 654)
Miscellaneous papers (1 vol) 1843-59, incl letters to T Martin. *Suffolk RO, Bury St Edmunds* (FL 586/13/1)
Papers and correspondence, incl diaries and 233 letters to Sir WJ Hooker 1826-57. *R Botanic Gardens, Kew*
Correspondence (61 items) with CR Darwin 1831-60. *Cambridge Univ Libr*
Letters (19) to Sir R Owen. *Brit Mus (Nat Hist) General Libr*
Letters (13) to W Hutton. *Tyne and Wear Archives Dept, Newcastle upon Tyne*
Correspondence (10 items) with Sir GB Airy 1835-*c*1856. *R Greenwich Observatory, Herstmonceux* (in MSS 938-46)

See also Darwin CR

[285] **HERON-ALLEN, Edward** (1861-1943)
FRS, zoologist

Drawings, papers and correspondence on the foraminifera. *Brit Mus (Nat Hist) Zoology Dept*

Letters from Florence Lockwood 1889-91. *Brit Libr* (Add MS 50507)

Correspondence (20ff) rel to a motto for the City of Westminster 1902. *Ibid* (Add MS 40166 H)

[286] **HERSCHEL, John** (1837-1921)
FRS, engineer and astronomer

Family and personal papers. *Private*

'History of pendulum observations', 'Account of observations' 1881-82, notes and bibliography on pendulums. *Royal Soc*

Correspondence (45 items) with Sir GG Stokes 1867-86. *Cambridge Univ Libr* (in MS Add 7656)

Letters (20) to officers of the Royal Astronomical Society 1874-94. *R Astronomical Soc*

[287] **HERSCHEL, Sir John Frederick William,** 1st Bt (1792-1871)
FRS, astronomer

Papers and correspondence, incl notes taken by him at lectures 1810-11, papers on mathematics (31), astronomy (31), the barometer (20), the actinometer (19), photography (19) and physics (14), star charts (12), memorandum books (8) 1825-32, 1850, 1852-53, papers rel to Greenwich Observatory, the Royal Mint and the Terrestrial Magnetism Survey, notes on chemistry (*c*2,900ff), account books (8) 1825-53, housekeeping books (3) 1822-40, diaries 1820-71, travel journals (12) 1809-50, and 2,555 items of correspondence 1814-70. *Texas Univ, Austin* [NRA 13677]

Correspondence (*c*10,000 items); drawings of nebulae. *Royal Soc* [NRA 8803]

Papers and correspondence, incl records of his observations at Slough, the Cape of Good Hope and Collingwood, Kent, 1816-71; MSS of his *Results*, *General catalogue of double stars*, *General history of double stars* and other published works, with related papers; 22 letters to officers of the Royal Astronomical Society 1820-68. *R Astronomical Soc*

Mathematical and astronomical papers, notes rel to Slough Observatory 1860, 4 notebooks 1810-12, and correspondence. *Houghton Libr, Harvard Univ, Cambridge, Mass* [NRA 19360]

Papers (15 vols), incl MSS of his published articles, unpublished paper on photography 1839, and catalogues of double stars. *St John's Coll, Cambridge* [NRA 9502]

Notebooks recording his chemical and photographic experiments, with related photographs. *Science Mus Libr*

Correspondence and papers mainly rel to his photographic work 1840-42. *History of Science Mus, Oxford* (MSS Mus 58, 113)

Geological notebook rel to the Isle of Wight 1831. *Inst of Geological Sciences*

Notes and correspondence rel to the chronometer and barometer 1832-33; household accounts kept by his wife 1832-38, 1852-86. *Nat Maritime Mus, Greenwich* [NRA 10522]

Family papers and correspondence, incl notes (1 vol) on astronomy, his translation of Homer etc *c*1852-66. *Private*

Papers and correspondence (48 items) mainly rel to education in Cape Colony *c*1833-38. *Cape Town Univ Libr*

Letters (2 vols) to him and his wife from Caroline Herschel 1822-46. *Brit Libr* (Egerton MSS 3761-62)

Letters (*c*130) to W Whewell 1817-66. *Trinity Coll, Cambridge* [NRA 8804]

Correspondence (119 items) with Sir GB Airy 1835-70. *R Greenwich Observatory, Herstmonceux* (in MSS 938-52)

Letters (50) to J Tyndall. *Royal Inst* [NRA 9522]

Letters (49) to Sir GG Stokes 1852-70. *Cambridge Univ Libr* (in MS Add 7656)

Letters (41) to Sir WR Hamilton 1833-65. *Trinity Coll, Dublin* (MS 1493) [NRA 20076]

Letters (39) to JD Forbes 1832-59. *St Andrews Univ Libr* [NRA 13132]

Letters (19) to Sir C Lyell 1836-68; correspondence (17 items) with WH Smyth 1827-38. *American Philosophical Soc, Philadelphia*

Letters (19) to Sir RI Murchison 1838-52. *Edinburgh Univ Libr* (in MSS Gen 523/4, 523/6)

Letters (14) to Sir WJ Hooker 1843-48. *R Botanic Gardens, Kew*

Correspondence (12 items) with C Babbage 1819-70. *Brit Libr* (Add MSS 37182-99 *passim*)

Poems and miscellaneous papers. *Cambridge Univ Libr* (MS 7617) [NRA 10522]

See also Airy, Andrews T, Babbage, Baily, Barlow, Beaufort, Bland, Brande, Brewster, Brodie BC the elder, Carpenter WB, Carrington, Clarke WB, Crookes, De La Rue, De Morgan, Faraday, Fitton, Forbes JD, Gassiot, Gilbert, Hooker WJ, Kater, Lindley, Lloyd H, Lloyd JA, Lubbock JW, Lyell, Maclear, Maskelyne, Murchison, Nasmyth, Owen, Peacock, Pole, Rigaud, Robinson TR, Sabine, Sedgwick, Sheepshanks, Smyth, Somerville, Stokes, Stratford, Sykes, Talbot, Tyndall, Wheatstone, Whewell, Wollaston, Young T

[288] **HERSCHEL, Sir William** (1738-1822)
FRS, astronomer

Papers and correspondence, incl records of his observations (55 vols) 1774-1819, catalogues of double stars, nebulae and clusters of stars, records of experiments on the construction of specula (4 vols) 1773-1818 and other papers rel to telescopes, autobiographical memoranda *c*1818, and *c*1,200 items of correspondence 1781-1821. *R Astronomical Soc*
Papers (73) submitted to the Royal Society 1780-1818, sweep books (7 vols) 1783-1802, observations and catalogues of nebulae and clusters of stars (2 vols and 5 bundles). *Royal Soc*
Journal, records of experiments on the construction of specula 1778-90 and other papers rel to telescopes. *History of Science Mus, Oxford* (MSS 117, 180)
Papers on astronomy (7 items) and physics (16 items), and some correspondence. *Texas Univ, Austin*
Papers, incl autobiographical memorandum 1784, and 31 papers read to the Bath Philosophical Society in 1781. *Private*
Memorandum of his journey to Göttingen 1786, catalogue of his scientific papers 1821, and list (1 vol) of his 291 correspondents. *Private*
Symphonies (9 vols). *Brit Libr* (Add MSS 49624-32)
Organ compositions and part of a treatise on the theory of music (6 vols). *Edinburgh Univ Libr* (MS Dk 7.35)
Musical compositions (16). *Music Libr, Univ of California, Berkeley*
Letters (25) to Sir J Banks 1781-1802. *Brit Mus (Nat Hist) Botany Dept*

See also Blagden, Hornsby, Maskelyne, Pond, Robertson A, Vince

[289] **HEWSON, William** (1739-1774)
FRS, surgeon and anatomist

Papers (7) read to the Royal Society 1768-74. *Royal Soc*
Student notes on his anatomy lectures *c*1770. *R Coll of Physicians, London*
Student notes (2 vols) on his anatomy lectures, with W Hunter, 1772. *R Coll of Surgeons, London*

[290] **HICKS, Henry** (1837-1899)
FRS, geologist

Notes on 'The oldest rocks in the British Isles' 1882. *Nat Libr of Wales* (MS 4440)

[291] **HIERN, William Philip** (1839-1925)
FRS, botanist and mathematician

Botanical papers. *Brit Mus (Nat Hist) Botany Dept*
Papers and correspondence, incl notes on central African plants collected by S Pinto *c*1880-89. *R Botanic Gardens, Kew*
MS on north Devon plants. *North Devon Athenaeum, Barnstaple*
Interleaved and annotated copy of his 'Monograph on ebenaceae', *Transactions of the Cambridge Philosophical Society*, xii (1873). *Linnean Soc*

[292] **HIGHMORE, Nathaniel** (1613-1685)
Physician

Medical correspondence and papers, incl 'Anatomia restaurata', 'Loci communes' 1676-83, medical and anatomical collections and observations. *Brit Libr* (Sloane MSS 534-82 *passim*, 1053)

[293] **HILL, Archibald Vivian,** CH (1886-1977)
FRS, physiologist

Correspondence and papers, incl diaries (57 vols) 1924-77, descriptions of visits to Russia 1935 and USA 1936, memoirs (3 vols), fellowship thesis on heat production in muscle 1910, 6 lectures on muscle and other unpublished writings, and over 10,000 items of correspondence 1908-77. *Churchill Coll, Cambridge*
Correspondence and papers rel to research topics etc. *Medical Research Council*

▷RG Frank, 'The AV Hill papers at Churchill College, Cambridge', *J Hist Biol*, 11, 1978, pp 211-14

See also Barcroft, Dale HH, Haldane, Hopkins, Sherrington

[294] **HILL, Sir Arthur William,** KCMG (1875-1941)
FRS, botanist

Papers, incl 12 diaries of his foreign travels *c*1902-37, zoological notebooks (2 vols), notebooks on Kew (3 vols), notes and drawings rel to orchidaceae 1892, list of plants collected in the Faroes and Iceland 1900, and material for his *Memoir* of the Revd HN Ellacombe (1919). *R Botanic Gardens, Kew*

[295] **HINDLE, Edward** (1886-1973)
FRS, biologist

Papers and correspondence *c*1905-66, incl laboratory and other notebooks, sketch books, travel journals, case notes, reports, lectures, MSS of articles, lectures and broadcasts. *Glasgow Univ Libr* [NRA 21908]
Correspondence with the Eugenics Society 1950-60. *Contemporary Medical Archives*

Centre, Wellcome Inst for the Hist of Medicine [NRA 24905]

See also Dale HH, Dobell

[296] **HINSHELWOOD, Sir Cyril Norman,** OM (1897-1967)
FRS, chemist

Papers 1919-43, incl laboratory notebooks and working papers on molecular and gas reactions, reports, graphs and notes on respirator design undertaken for the Chemical Defence Board during World War II. *Royal Soc* [NRA 18546]

[297] **HIRST, Thomas Archer** (1830-1892)
FRS, mathematician

Journals (5 vols) as student in Marburg and Berlin 1850-53. *Royal Inst* (Tyndall Papers)
Letters (*c*540) to him 1853-92. *University Coll London* (London Mathematical Society Papers)
Letters (15) to officers of the Royal Astronomical Society 1866-84. *R Astronomical Soc*

See also Cayley A, Salmon, Sylvester

[298] **HODGSON, Brian Houghton** (1800-1894)
FRS, naturalist

Correspondence and papers rel to his published works on Nepal down to 1864, incl papers on its topography, ethnography, languages, religion, agriculture and economics. *India Office Libr and Records* (Hodgson MSS)
Correspondence and papers, incl ornithological notebooks, zoological diary of journey from Monghyr to Darjeeling 1846, MS on the anatomy and habits of quadrupeds and birds in Nepal 1833-40, drawings (9 vols) of mammals and birds of India and Nepal. *Zoological Soc of London*
Correspondence and papers, incl descriptions (2 vols) of quadrupeds and birds, MSS submitted by him to learned societies, mostly on Indian zoology, correspondence (2 vols) with JE Gray and others *c*1830-45, correspondence with J Gould and others concerning a projected work on the zoology of Nepal 1835-37, and drawings (1319 in 7 vols) of the vertebrates of Nepal. *Brit Mus (Nat Hist) Zoology Dept*
Correspondence and papers 1819-94. *Bodleian Libr, Oxford* (MSS Hodgson 9-17; MS Eng hist c 262) [NRA 17019]
Correspondence and papers (16 packets), incl sketches 1820-43. *R Asiatic Soc*
Catalogue of Nepalese mammals 1838. *Linnean Soc*
Catalogue of Nepalese birds 1845. *R Coll of Surgeons, London*
Correspondence with the Earl of Ellenborough 1842-44. *Public Record Office* (PRO 30/12/55, 88) [NRA 21870]
Letters (12) to Sir WJ Hooker 1853-62. *R Botanic Gardens, Kew*
Letters to ACLG Günther 1861-67. *Brit Mus (Nat Hist) General Libr*

[299] **HOME, Sir Everard,** 1st Bt (1756-1832)
FRS, surgeon

Papers, incl catalogue of museum specimens, chiefly anatomical, additions and alterations to W Clift's catalogue of the physiological series of the Hunterian Collection, MSS and annotated copies of articles published in *Philosophical Transactions* (1807-11), notes for his lectures and notes taken by students 1794-1814, and copies of his letters to his son 1810-22. *R Coll of Surgeons, London*
Papers and correspondence, incl texts of lectures 1810-13 published in his *Lectures on comparative anatomy*, outline of a physiology catalogue for the Royal College of Surgeons, and a volume of natural history letters and papers. *Brit Mus (Nat Hist) General Libr*
Texts of 4 lectures on physiology and anatomy 1810 published in his *Lectures on comparative anatomy. Brit Libr* (in Add MS 34407)
MSS of 90 papers published in *Philosophical Transactions* and 30 other papers submitted to the Royal Society. *Royal Soc*
Lectures (5 vols) on botany 1822. *Cambridge Univ Libr* (MSS Add 6213-17)
Correspondence with Baron de Cuvier. *Bibl, Institut de France, Paris*
Letters (10) to Sir J Banks 1783-1804. *R Botanic Gardens, Kew*

See also Clift

[300] **HOOKE, Robert** (1635-1703)
FRS, natural philosopher

Papers and correspondence, incl essay on the inflection of a direct motion into a curve 1666, discourse concerning Newton's theory of light, account of Burnet's 'Archaeologiae philosophicae', method for making a history of the weather, and drawings. *Brit Libr* (Sloane MSS 698, 917, 1039, 1676-77, 1942, 3823, 4062, 4067; Add MSS 5238, 6193-6209)
Papers 1672-81, incl diary, travel journals, reports as surveyor for rebuilding London after the Great Fire, anatomical observations, lectures, and therapeutical notes. *Guildhall Libr, London*
Papers (97) and correspondence (20 items) 1677-85. *Royal Soc*
Papers, incl 'Philosophicall scribbles', and some correspondence. *Trinity Coll, Cambridge*

▷GL Keynes, *Bibliography of Dr Robert Hooke*, 1960

[301] **HOOKER, Sir Joseph Dalton,** OM, GCSI (1817-1911)
FRS, botanist

Correspondence (*c*55 vols) and papers, incl *c*50 journals and notebooks kept on Antarctic, Indian and other travels 1839-79, floras, plant lists, drawings and diplomas. *R Botanic Gardens, Kew*
Lists, notes and drawings of Antarctic plants collected 1839-43; drawings of Indian fungi. *Brit Mus (Nat Hist) Botany Dept*
Drawings (28) of Antarctic fish *c*1839-43. *Ibid, Zoology Dept*
Drawings of invertebrates 1840-41. *Ibid, Mineralogy Dept*
Miscellaneous correspondence and papers, incl a flora of Prince of Wales Island and papers rel to the *Index Kewensis* 1881-1906. *Linnean Soc*
Correspondence (5 vols) with CR Darwin 1843-82, and annotated draft of part of Darwin's *Origin of species* (1859). *Cambridge Univ Libr* (DAR 15, 100-04) [NRA 11458]
Correspondence (*c*430 items) with TH Huxley 1853-95. *Imperial Coll, London*
Letters (200) to A Gray 1844-87. *Arnold Arboretum and Gray Herbarium Libr, Harvard Univ, Cambridge, Mass*
Letters (*c*200) to J Tyndall. *Royal Inst*
Correspondence (2 vols) with T Anderson *c*1860-69. *R Botanic Garden, Edinburgh*
Letters (92) to Sir FJH von Müller 1857-76. *R Botanic Gardens and National Herbarium, South Yarra, Australia*
Letters (41) to Sir GG Stokes 1869-89. *Cambridge Univ Libr* (in MS Add 7656)
Letters (11) to officers of the Royal Geographical Society 1855-78, and letters to J Hogg describing his travels in Tibet 1849-50. *R Geographical Soc*
Letters (10) to W Gourlie 1847-55. *Nat Libr of Scotland* (in MS 10789)
Letters to ACLG Günther 1861-1909. *Brit Mus (Nat Hist) General Libr*

See also Darwin CR, Forbes E, Thiselton-Dyer, Tyndall, Vines

[302] **HOOKER, Sir William Jackson** (1785-1865)
FRS, botanist

Correspondence (85 vols) and papers, incl an account of the natural history of Yarmouth and its environs 1807-40, journal of a tour in Switzerland 1814, lectures on botany (3 vols), catalogue of his herbarium 1822, catalogues of his library *c*1816-63, catalogue and transcripts made *c*1810-12 of W Roxburgh's 'Flora Indica' (10 vols), notes, drawings and diplomas. *R Botanic Gardens, Kew*
Notes and drawings of his copy of NL Burmann's *Flora Indica* (1768). *Brit Mus (Nat Hist) Botany Dept*
Lists and identifications inserted in HS Tiffen's album 'New Zealand Company: native plants of New Zealand' *c*1840-42. *R Commonwealth Soc*
Letters (150) to A Gray 1834-65. *Arnold Arboretum and Gray Herbarium Libr, Harvard Univ, Cambridge, Mass*
Letters (70) from J Jörgensen 1809-25. *Brit Libr* (in Egerton MS 2070)
Letters (*c*42) to Sir WC Trevelyan *c*1861-65. *Newcastle Univ Libr* [NRA 12238]
Letters from him, incl 23 to W Swainson 1816-39 and 13 to Sir JE Smith 1813-26. *Linnean Soc*
Letters (26) to officers of the Royal Geographical Society 1842-62. *R Geographical Soc*
Correspondence (17 items) with Sir JFW Herschel 1843-58. *Royal Soc*

▷*Annals of Botany*, xvi, 1902, App C, pp clxxxvii-ccxx

See also Arnott, Babington, Ball J, Beaufort, Bennett, Bentham, Berkeley MJ, Bloxam, Boott, Borrer, Bromfield, Brown R, Buckland, Burchell, Daubeny, Falconer, Fielding, Forster, Gardner, Gray, Greville, Griffith, Hanbury, Henslow, Herschel JFW, Hodgson, Jardine, King PP, Lambert, Lindley J, Lowe, Lyell, MacGillivray J, Miers J, Newton A, Owen, Sabine, Saunders, Seemann, Spruce, Swainson, Sykes, Talbot, Thwaites, Trevelyan, Turner D, Wallich N, Watson, Wilson, Winch

[303] **HOPE, Revd Frederick William** (1797-1862)
Entomologist

Entomological notebooks and correspondence. *Hope Libr, University Mus, Oxford*
Register (3 vols) of living and fossil molluscs and brachiopods, and drawings (2 packets) of fossil plants. *University Mus, Oxford* (Geological Collns)

[304] **HOPE, John** (1725-1786)
FRS, botanist and physician

Correspondence and papers, incl meteorological observations 1758, notes on his Highland journey 1765, notes on the proposed Scottish College of Physicians 1784, medical case notes and letters from physicians, notes for and ıel to his botany lectures, correspondence with botanists, and correspondence and papers rel to the botanic gardens in Edinburgh. *Scottish Record Office* (GD 253/144-45) [NRA 13863]

Clinical case book, Royal Infirmary, Edinburgh, 1781. *Edinburgh Univ Libr* (MS Dc 8.169)
Notes and drawings for his botany lectures, with student notes on them 1777-83, and annotations on a list of Edinburgh plants 1765 and on a catalogue of British plants in his herbarium 1768. *R Botanic Garden, Edinburgh*
Papers (3) submitted to the Royal Society. *Royal Soc*
Correspondence (21 items) with Sir J Banks 1766-86. *R Botanic Gardens, Kew*
Correspondence with R Pulteney 1763-86. *Linnean Soc*

[305] **HOPE, Thomas Charles** (1766-1844)
FRS, chemist

Notes (35 vols) taken by him of lectures by J Black, W Cullen, James Gregory, Alexander Monro II, J Robison and others *c*1782-87; notes for his own chemistry lectures and other papers *c*1790-1842. *Edinburgh Univ Libr* (MSS Dc 10.9-15; MSS Gen 268-72)
Clinical lectures and case notes, Royal Infirmary, Edinburgh, 1796-98. *R Coll of Physicians, London*
Clinical casebook 1796-97. *R Coll of Physicians, Edinburgh* [NRA 16015]
Letters (13) to Sir JE Smith 1784-94. *Linnean Soc*

[306] **HOPKINS, Sir Frederick Gowland,** OM (1861-1947)
FRS, biochemist

Papers and correspondence, incl papers on a scheme for nutritional studies at Cambridge 1927-28, research reports to the Medical Research Council 1933-36, and *c*250 items of correspondence 1914-47. *Medical Research Council*
Miscellaneous personal papers (1 box). *Cambridge Univ Libr* (MS Add 7620)
Correspondence (over 20 items) with AV Hill. *Churchill Coll, Cambridge*

[307] **HORNER, Leonard** (1785-1864)
FRS, geologist

Correspondence and papers, incl geological notes, notes taken on his journey to Ems 1830, journal of his visit to Florence 1861-62, and diplomas 1814-46. *Nat Libr of Scotland* (MSS 2213-22, Ch 934-42)
Papers and correspondence, incl notes on the Huttonian system of geology 1809, sketch of the geology of south-west Somerset 1815, description, with GB Greenough and R Phillips, of fuller's earth pits at Nutfield, and 21 letters to Sir RI Murchison. *Geological Soc of London*
Sketch (3 vols) of the geology of west Somerset 1828. *Somerset RO, Taunton*
Accounts of geological features and specimens *c*1810-40. *Inst of Geological Sciences* [NRA 18675]
Letters (42) to Lord Brougham 1824-59. *University Coll London*
Letters (33) to AJG Marcet 1810-22. *Nat Libr of Scotland* (in MS 9818)
Letters (*c*31) rel to college affairs 1827-42. *University Coll London*
Letters (30) to J Loch 1827-32. *Ibid* (in MS Add 131)
Correspondence (20 items) with A De Morgan 1828-31. *London Univ Libr* (in MS 322)
Correspondence (15 items) with Sir R Peel 1824-45. *Brit Libr* (in Add MSS 40369-40567)
Letters (11) to the Society for the Diffusion of Useful Knowledge 1830-35. *University Coll London*

[308] **HORNSBY, Thomas** (1733-1810)
FRS, astronomer

Records of transit observations (8 vols) 1774-1805, zenith distance observations (4 vols) 1773-1803, and zenith sector observations; correspondence (10 items) with Sir W Herschel 1774-87. *R Astronomical Soc*
Astronomical and mathematical papers, incl astronomical computations, observations and tables, with copies of older observations, 1689-1800, catalogue of fixed stars *c*1768, his corrected proofs of his edition of Bradley's *Astronomical observations*, I (1798), and extracts from his letters to the Duke of Marlborough *c*1784-99. *History of Science Mus, Oxford* (MSS Radcliffe 1-40, 54, 67, 71-73)
Astronomical papers and notes *c*1768-74, incl transit and quadrant observations 1770-73 and catalogue of circumpolar stars. *Corpus Christi Coll, Oxford* (MSS 512-14)
Lectures on natural philosophy. *Bodleian Libr, Oxford* (MSS Rigaud 54, Radcliffe dep d 1-15, Radcliffe Trust d 9-23, e 11-12, f 4)
Notes and miscellaneous papers. *Ibid* (MSS Radcliffe dep e 1-2, f 1, Top Oxon c 236, Eng lett c 132)
Letters (81ff) to F Wingrave 1789-95. *Lambeth Palace Libr* (MS 1802)

[309] **HORSLEY, Revd Samuel** (1733-1806)
FRS, mathematician

Papers on mathematics, astronomy, theology, Roman and biblical chronology, Greek literature, notes on Sir I Newton and his *Historia universalis*, annotated tracts on mathematics and physics. *Royal Soc*
Correspondence 1769-70, 1785-1806, and transcripts of his writings. *Lambeth Palace Libr* (MSS 1767, 1890)

[310] HOWARD, Luke (1772-1864)
FRS, meteorologist and chemist

Correspondence and papers, incl diary 1807, journals and cash accounts (9 vols) 1811-19, paper on the advantages of civilisation 1829, and early business records of Howard & Sons Ltd, chemical manufacturers. *Greater London RO* (Accs 1017, 1037, 1270) [NRA 14651]

Miscellaneous papers and correspondence (13 items) 1806-37. *Soc of Friends Libr, London* [NRA 20017]

[311] HUDSON, Charles Thomas (1828-1903)
FRS, naturalist

Original drawings for his *Rotifera* (1886-89). *Royal Soc*

[312] HUGGINS, Sir William, OM, KCB (1824-1910)
FRS, astronomer

Correspondence and papers, incl 12 sheets of drawings of Jupiter, 285 letters to officers of the Royal Astronomical Society 1854-1900 and 11 concerning the solar eclipse of 1870. *R Astronomical Soc*

Correspondence (325 items) 1865-1908. *Royal Soc*

Letters (186) to Sir GG Stokes 1864-99, and 23 to the Revd TR Robinson 1866-74. *Cambridge Univ Libr* (in MS Add 7656)

Letters (15) to Sir AB Kempe 1894-1905. *West Sussex RO, Chichester*

Correspondence with GE Hale 1891-1910. *California Inst of Technology, Pasadena*

Letters (25ff) to Sir HW Acland 1866-92. *Bodleian Libr, Oxford* (in MSS Acland d 63, 83)

[313] HUGHES, David Edward (1830-1900)
FRS, electrician

Records (10 vols) of his experiments rel to telegraphy and the microphone 1860-86; texts (1 vol) of his lectures to scientific societies 1883-86. *Brit Libr* (Add MSS 40161-63, 40641-48)

[314] HUME-ROTHERY, William (1899-1968)
FRS, metallurgist

Correspondence and papers 1922-68, incl laboratory notebooks 1944-68 and other working papers, and texts of his lectures *c*1957-1967. *Bodleian Libr, Oxford* [NRA 17174]

Correspondence (11 files) mainly rel to his publications. *Metallurgy and Science of Materials Dept, Oxford*

Correspondence with Viscount Cherwell 1928-43. *Nuffield Coll, Oxford* [NRA 16447]

Correspondence with EC Stoner 1940-55. *Brotherton Libr, Leeds Univ* (MS 333) [NRA 17735]

[315] HUMPHREY, Herbert Alfred (1868-1951)
Engineer

Papers and correspondence 1883-1948, incl his student notebooks, working papers, reports, patent specifications and engineering drawings. *Imperial Coll, London* [NRA 14865]

[316] HUNTER, John (1728-1793)
FRS, surgeon and anatomist

Papers and correspondence, incl anatomical observations, MSS of his *Natural history of the human teeth*, pt i (1771), *Treatise on the venereal disease* (1786), and part of his *Observations and reflections on geology* (1859), transcripts of his writings (originals destroyed), 32 letters to E Jenner 1773-93, and student notes (14 sets) of his lectures 1786-91. *R Coll of Surgeons, London* [NRA 9521]

Papers submitted to the Royal Society. *Royal Soc*

Anatomical notes (5ff), published in *Essays and observations*, ed Sir R Owen (1861). *Brit Libr* (in Add MS 34407)

Letters addressed to him, and notes. *Brit Mus (Nat Hist) General Libr*

▷ JB Bailey, *Catalogue of the collection of Hunterian relics exhibited at the Royal College of Surgeons of England*, 1893

See also Clift

[317] HUNTER, William (1718-1783)
FRS, anatomist

Correspondence and papers, incl a journal of his attendance on Queen Charlotte 1762-65, pharmaceutical notebooks (6 vols), notes on drawings of calculi (3 vols) *c*1764-1778, list of books lent 1771-81, unpublished work on the lues venerea 1775, MS of his *Anatomical description of the human gravid uterus* (1794), and student notes of his lectures. *Glasgow Univ Libr* (Hunterian MSS *passim*)

Miscellaneous papers incl his diplomas, and student notes (17 sets) of his lectures 1755-*c*1783. *R Coll of Surgeons, London* [NRA 9521]

His notes (2 vols) for lectures, and student notes of his lectures. *St George's Hospital Medical School Libr, London*

Papers submitted to the Royal Society. *Royal Soc*

Paper on aneurisms. *Bodleian Libr, Oxford* (in MS Eng misc e 29)

See also Hewson

[318] **HURST, Charles Chamberlain** (1870-1947)
Geneticist

Correspondence, notebooks and other papers. *Cambridge Univ Libr*

See also Bateson

[319] **HUTCHINSON, John** (1884-1972)
FRS, botanist

Correspondence and papers, incl botanical notes, plant lists, account with TA Sprague of botanical expedition to the Canary Islands 1913, account of a botanical journey round the world 1960-61, memoir (2 vols) 'From potting shed to FRS', MS of his *Families of flowering plants*, 3rd edn (1973), and drawings for his published works. *R Botanic Gardens, Kew*

[320] **HUTTON, James** (1726-1797)
Geologist

MS of his *Theory of the earth*, III (1899). *Geological Soc of London*

Unpublished work (2 vols) on the principles of agriculture. *Royal Soc of Edinburgh*

'Memorial justifying the present theory of the earth from the suspicion of impiety', and 5 letters. *Fitzwilliam Mus, Cambridge* (SG Perceval Colln)

Sketches (50) for his *Theory of the earth*. *Private*

[321] **HUXLEY, Sir Julian Sorell** (1887-1975)
FRS, biologist

Correspondence and papers *c*1895-1975, incl diaries, notebooks, sketch books, material for and MSS of his published and unpublished works. *Woodson Research Center, Rice Univ Libr, Houston*

Ornithological diaries, notes and MSS 1903-49. *Edward Grey Inst, Zoology Dept, Oxford*

MS, with FS Callow, on the asymmetry of male fiddler crabs, and some correspondence. *Zoology Dept Libr, Oxford*

Correspondence (2 files) with the Eugenics Society 1928-69. *Contemporary Medical Archives Centre, Wellcome Inst for the Hist of Medicine* [NRA 24905]

Correspondence (32 items) with Marie Stopes 1924-53. *Brit Libr* (in Add MS 58473)

Correspondence (28ff) with Gilbert Murray 1924-53. *Bodleian Libr, Oxford* [NRA 16865]

Letters (17) to HT Clarke 1958-72. *American Philosophical Soc, Philadelphia*

Correspondence (14ff) with A Creech Jones 1944-63. *Rhodes House Libr, Oxford* (in MSS Brit Emp s 332/Box 7/2) [NRA 14026]

Correspondence with the Marquess of Lothian 1928-39. *Scottish Record Office* (in GD 40/17) [NRA 10737]

Correspondence with Viscount Cherwell 1940. *Nuffield Coll, Oxford* [NRA 16447]

See also Bateson

[322] **HUXLEY, Thomas Henry** (1825-1895)
FRS, man of science

Correspondence and papers (*c*270 vols and 8 boxes), incl notes and sketches (5 vols) made during the voyage of HMS *Rattlesnake* 1847-49, pocket diaries (38) 1857-94, notebooks (*c*160) mainly rel to zoology, botany and his lectures on biology 1840-94, sketch books, drawings, photographs, correspondence with his wife 1847-95, other family correspondence and *c* 4,500 items of scientific correspondence. *Imperial Coll, London* [NRA 8194]

Letters (*c*300) to Sir J Tyndall and miscellaneous letters and papers (63 items). *Royal Inst*

Letters (184) to Sir M Foster 1875-95. *R Coll of Physicians, London*

Letters (84) to Macmillan & Co 1866-94. *Brit Libr* (Add MS 55210)

Correspondence (81 items) with Sir GG Stokes 1864-93. *Cambridge Univ Libr* (in MS Add 7656)

Letters (*c*50), incl 11 to Sir H Cole 1857-87 and 9 to Sir C Lyell 1859-68. *American Philosophical Soc, Philadelphia*

Correspondence (38 items) with the Revd G Gordon 1857-78. *Private* [NRA 22745]

Letters (15) to T Reeks 1856-79. *Inst of Geological Sciences* [NRA 18675]

Correspondence (12 items) with CR Darwin 1859-80, and MS of part of his chapter on the reception of his *Origin of species* published in *Life and letters of Charles Darwin*, II (1887). *Cambridge Univ Libr*

Letters (11) to Lord Playfair 1885-91. *Imperial Coll, London* [NRA 11556]

▷ WR Dawson, *The Huxley papers. A descriptive catalogue of the correspondence, manuscripts and miscellaneous papers preserved in the Imperial College*, 1946

See also Allman, Armstrong WG, Bateson, Carpenter PH, Carpenter WB, Darwin CR, Darwin F, Flower, Forbes E, Foster M, Frankland, Hooker JD, Lankester, Lockyer, Lubbock J, Lyell, Mivart, Murchison, Paget, Playfair, Rolleston, Roscoe, Spottiswoode, Thiselton-Dyer, Thomson CW, Tyndall

[323] **IRVINE, Sir James Colquhoun,** KBE (1877-1952)
FRS, chemist

Papers, incl diaries, journals and experimental observations. *St Andrews Univ Libr and Chemistry Dept*

[324] **IVORY, Sir James** (1765-1842)
FRS, mathematician

Papers (26 items) rel to him and to his nephew Lord Ivory 1814-40. *Scottish Record Office* (GD 1/547/3-28)
Letters (25) to C Babbage 1815-20. *Brit Libr* (in Add MSS 37182, 37185-86, 37188, 37200)
Letters (13) to Sir JW Lubbock 1830-40. *Royal Soc*

[325] **JACKSON, Sir Herbert,** KBE (1863-1936)
FRS, chemist

Correspondence rel to college administration 1892-1920. *King's Coll, London*
Correspondence with Viscount Cherwell on the quartz lens 1915 and X-ray glass apparatus 1916. *Nuffield Coll, Oxford* [NRA 16447]
Correspondence with GE Hale 1929-33. *California Inst of Technology, Pasadena*

[326] **JACKSON, Willis,**
Baron Jackson of Burnley (1904-1970)
FRS, electrical engineer

Lectures, articles, notebooks, diary, working papers and correspondence *c*1922-1970; scrapbooks (91 vols) containing reports, newspaper cuttings, photographs, papers and correspondence 1916-70. *Imperial Coll, London* [NRA 18379]

See also Blackett

[327] **JAMESON, Robert** (1774-1854)
FRS, mineralogist

Correspondence and papers, incl journal of a voyage from Leith to London 1793; notebooks (13) containing diaries of tours, mineralogical notes of walks, summary of correspondence etc 1793-1819; geological and botanical notes; student notes on his lectures 1813-36; some family papers. *Edinburgh Univ Libr* (in MSS Gen 121-31, MSS Dc 1-10 *passim*, MS Dk 5. 23, Pollok-Morris MSS)
Catalogue of the minerals in the Hunterian Museum 1809. *Glasgow Univ Libr* (in Hunterian MSS)
Correspondence (31 items) with JD Forbes 1832-51. *St Andrews Univ Libr* [NRA 13132]
Letters (28) to GB Greenough 1811-33. *University Coll London*

[328] **JARDINE, Sir William,** 7th Bt (1800-1874)
Naturalist

Correspondence and papers, incl a journal of his travels in Holland 1825, botanical and zoological notes and drawings, drafts of articles, introduction and drawings for *Contributions to ornithology*, scrapbook containing notes, maps, letters etc rel to the Salmon Fisheries Survey 1860, and 9 boxes of correspondence 1821-48. *R Scottish Mus, Edinburgh*
Correspondence and papers, mainly rel to ornithology, incl material for the *Naturalist's Library*, annotated MS of his edition of A Wilson's *American ornithology* (1832), and letters from J Gould. *Brit Mus (Nat Hist) Zoology Dept*
Journals of his travels in Ireland 1845 and the W Indies 1860-61, calendar of ornithology for 1849, and 315 letters to him 1828-41. *Edinburgh Univ Libr* (MSS Dk 6. 20-21)
Miscellaneous letters (1 vol) addressed to him 1852. *Brit Mus (Nat Hist) General Libr*
Letters (22) to TC Eyton 1838-71. *American Philosophical Soc, Philadelphia* [NRA 9728]
Letters (14) to W Swainson 1827-31. *Linnean Soc* [NRA 22111]
Letters (14) to Sir WJ Hooker 1830-51. *R Botanic Gardens, Kew*
Correspondence with PJ Selby. *Mineralogy and Petrology Dept, Cambridge*

See also Forbes E, Gould, Gray, Greville, Watson

[329] **JEANS, Sir James Hopwood,** OM (1877-1946)
FRS, physicist and mathematician

Correspondence and papers. *Royal Soc*
Paper, with HT Flint and Sir OW Richardson, on the number of chemical elements, and a new uncertainty principle; letters (23) to Richardson 1906-34. *Texas Univ, Austin* [NRA 22337]
Correspondence with Viscount Cherwell 1916-*c*1931. *Nuffield Coll, Oxford* [NRA 16447]
Correspondence with GE Hale 1917-32. *California Inst of Technology, Pasadena*

[330] **JEFFREYS, John Gwyn** (1809-1885)
FRS, conchologist

Correspondence, sketches, plates and papers rel to molluscs 1859-83. *Smithsonian Inst, Washington*
His annotated copy (5 vols) of *British conchology* (1862-69). *Radcliffe Science Libr, Oxford*
Letters (61) to OAL Mörch 1862-77. *Brit Mus (Nat Hist) General Libr*
Letters to the Revd G Gordon *c*1860-70. *Private* [NRA 22745]

[331] **JENNER, Edward** (1749-1823)
FRS, physician

Papers and correspondence, incl diaries of visits to patients 1794-1803, accounts 1770-1823, prescriptions 1823, essay on marriage 1783, poems (2 vols), drafts annotated by him of his *Inquiry into . . . the cow pox* 1797 and 'An Act for preventing the spread of the small pox' 1808, and 74 letters to T Pruen 1808-22. *Wellcome Hist Medical Libr* (MSS 2068, 3014-28, 3072)

Diary containing observations on the cuckoo and notes of dissections of birds and domestic animals 1787-1806; papers submitted to the Gloucestershire Medical Society 1788-90; essay on nosology; miscellaneous correspondence (11 items) 1787-1821. *R Coll of Physicians, London*

MS of his *Inquiry into . . . the cow pox* 1797; 36 letters from him. *R Coll of Surgeons, London* [NRA 9521]

Diary notes, poems, and 23 letters to the Hodges family of Arlingham 1782-1823. *Gloucestershire RO* (D 18/659) [NRA 9191]

'Report of a case of hydatids of the kidney' 1790. *Osler Libr, McGill Univ, Montreal*

Observations on the cuckoo 1787. *Royal Soc*

Papers (64 items) 1800-22. *Duke Univ Medical Libr, Durham, N Carolina*

Correspondence (101 items), incl 21 letters to AJG Marcet. *William H Welch Medical Libr, Johns Hopkins Univ, Baltimore*

Letters (18) to AJG Marcet. *Royal Soc of Medicine, London*

Volume of transcripts by his nephew (originals partly destroyed) of his notes on vaccination, prescriptions, diary entries, weather records, quotations, reflections, answers to an insurance questionnaire etc 1788-1823. *William L Clements Libr, Univ of Michigan, Ann Arbor*

▷WR LeFanu, *A bio-bibliography of Edward Jenner, 1749-1823*, 1951

See also Hunter J

[332] **JERVIS-SMITH, Revd Frederick John** (1848-1911)
FRS, engineer and inventor

Correspondence and papers, incl patents granted to him, with specifications and related papers, 1881-1906, and scientific correspondence 1883-1911. *History of Science Mus, Oxford* (MSS Mus 29, 62, 89) [NRA 9532]

Letters to WH Preece 1899, enclosing notes on his experiments with microphones in 1877. *Inst of Electrical Engineers* (NAEST 21) [NRA 20573]

See also Berkeley RMTR

[333] **JOHNSON, Manuel John** (1805-1859)
FRS, astronomer

Contribution to a volume of 'Original entries of the comparisons of various standard scales' 1834-36, 4 papers submitted to the Royal Astronomical Society 1832, 1834 and nd, 14 letters to its officers 1843-59 and 24 to the Revd R Sheepshanks 1839-54. *R Astronomical Soc*

MS of his prize memoir on Bradley's Kew and Wanstead observatories, introduction to a catalogue of the principal stars in the southern hemisphere, notes rel to the East India Company's observatory at St Helena, and astronomical observations. *History of Science Mus, Oxford* (MSS Radcliffe 49-50) [NRA 9532]

Correspondence (12 items) with Sir GB Airy 1849-59. *R Greenwich Observatory, Herstmonceux* (in MSS 941-48)

[334] **JOHNSON, Revd Samuel Jenkins** (1845-1905)
FRS, astronomer

Projections of solar and lunar eclipses 1800-2000, projections of eclipses visible in England 538-2500, and 359 letters to officers of the Royal Astronomical Society 1871-1900. *R Astronomical Soc*

Projections of eclipses 1882-2000. *Royal Soc*

[335] **JONES, Sir Harold Spencer,** KBE (1890-1960)
FRS, astronomer

Correspondence and papers as astronomer royal 1933-55, incl papers rel to the continuation of the Royal Observatory's work during World War II and its subsequent move from Greenwich to Herstmonceux. *R Greenwich Observatory, Herstmonceux*

Diplomas and certificates (*c*20), and letters (*c*20) of appointment and congratulation, 1906-55. *Nat Maritime Mus, Greenwich*

[336] **JONES, Thomas Rupert** (1819-1911)
FRS, geologist and palaeontologist

Catalogue of fossil collections; annotated drawings (1 vol) of fossil ostracoda. *Brit Mus (Nat Hist) Palaeontology Dept*

[337] **JOULE, James Prescott** (1818-1889)
FRS, physicist

Papers and correspondence, incl laboratory notebook 1843-58, index of scientific papers, personal dictionary of scientific information, and 30 items of correspondence 1841-87. *Univ of Manchester Inst of Science and Technology* [NRA 9527]

MSS of papers published in *Philosophical Transactions*; miscellaneous letters (18). *Royal Soc*
Correspondence (244 items) with Lord Kelvin 1847-82. *Cambridge Univ Libr* (in MS Add 7342)
Correspondence (103 items) with Lord Kelvin 1853-73. *Glasgow Univ Libr*
Letters (36) to Lord Playfair 1844-48, and papers rel to their joint research on atomic volumes and specific gravity. *Manchester Literary and Philosophical Soc*
Letters (17) to Sir GG Stokes 1847-77. *Cambridge Univ Libr* (in MS Add 7656)

[338] **JOURDAIN, Revd Francis Charles Robert** (1865-1940)
Zoologist

Ornithological and zoological papers 1876-1937, incl records of weights and measurements of birds and eggs, notebook on coleoptera, ornithological bibliographies by regions 1911, 1916, MSS of his published works, and extracts from the published and unpublished works of others. *Edward Grey Inst, Zoology Dept, Oxford*
Entomological notebooks. *Hope Libr, University Mus, Oxford*

[339] **KATER, Henry** (1777-1835)
FRS, man of science

Notebook; correspondence (34 items) with Sir JFW Herschel 1822-32. *Royal Soc*
Report to the Central Board of Health on the propagation of cholera spasmodica 1831. *R Coll of Surgeons, London*
Paper submitted to the Royal Astronomical Society 1833. *R Astronomical Soc*

[340] **KEILIN, David** (1887-1963)
FRS, biologist

Papers and correspondence 1908-63, incl working notebooks and papers, lectures, MSS of his published works and drawings. *Cambridge Univ Libr* [NRA 17358]
MSS and drawings, with GHF Nuttall, on the human louse, 1919, 1930 and nd. *Zoology Dept Libr, Oxford*
Miscellaneous correspondence. *Medical Research Council*

[341] **KEIR, James** (1735-1820)
FRS, chemist

Correspondence (108 items) with M Boulton and others, and miscellaneous accounts etc, *c*1772-1816. *Birmingham Reference Libr* (Assay Office Boulton Colln, Boulton & Watt Colln and Muirhead Colln) [NRA 9497]
Papers (5) submitted to the Royal Society. *Royal Soc*

[342] **KEITH, Sir Arthur** (1866-1955)
FRS, anatomist

Correspondence and papers, incl diaries (4 vols) 1908-33, MS of his *Darwin revalued* (1955), and bibliography of his other writings. *R Coll of Surgeons, London*
Miscellaneous correspondence rel to research topics. *Medical Research Council*
Miscellaneous letters (12). *Royal Inst*

KELVIN, Baron, see Thomson, W

[343] **KENDALL, Percy Fry** (1856-1936)
Geologist

Notebooks (95) containing field observations and extracts from published works on stratigraphy, water supply etc, 1882-1930; geological notes on the crag deposits of E Anglia. *Inst of Geological Sciences* [NRA 18675]
Letters (113) addressed to him 1886-1932. *Brotherton Libr, Leeds Univ* (MS 570) [NRA 24897]

[344] **KERR, Sir John Graham** (1869-1957)
FRS, zoologist

Correspondence and papers 1886-1957, incl notes taken by him at zoology lectures 1886-*c*1896, and correspondence and papers rel to his S American expedition 1896-97, the Scottish Marine Biological Association 1901-16, war camouflage 1914-50, and the proposed Mid Scotland Ship Canal mainly 1942-45. *Glasgow Univ Archives* [NRA 14019]

[345] **KING, Sir Edmund** (1629-1709)
FRS, physician

Medical papers and collections 1664-96; letters (6) to Sir H Sloane 1701. *Brit Libr* (in Sloane MSS 1586-91, 1593-94, 1597-98, 1640, 4038, 4078)
Papers (10) submitted to the Royal Society. *Royal Soc*

[346] **KING, Philip Parker** (1793-1856)
FRS, naturalist and meteorologist

Correspondence and papers, incl New Zealand diary 1826, natural history notes made on voyages of HMS *Adventure* and HMS *Beagle* 1826-30, register of letters written 1832-40, field book 1837-38, notebook 1837-43, journal of his visit to New Zealand and Norfolk Island in HMS *Pelorus* 1838-39, account book 1848-56, and scientific and family correspondence. *Mitchell Libr, Sydney* (A 3599, MSS 592, 673, 963)

Journal and log books kept as officer commanding HMS *Mermaid* 1817-21 and HMS *Bathurst* 1821-22. *Naval Hist Libr, London* (MSS 72, 73/2, 79)
Correspondence rel to Parramatta Observatory 1847-55. *Nat Libr of Australia, Canberra* (Nan Kivell 5213)
Letters (20) to Sir WJ Hooker 1840-52. *R Botanic Gardens, Kew*

[347] **KIRBY, Revd William** (1759-1850)
FRS, entomologist

Catalogue of the British insects in his collection at Barham; catalogue (3 vols) of British staphylinidae. *Brit Mus (Nat Hist) Entomology Dept*
Diary of journeys in Suffolk and Norfolk 1802-08. *Suffolk RO, Ipswich* (HA 21/C1/1) [NRA 11873]
Entomological paper, diplomas and some correspondence *c*1791-1829. *Linnean Soc*

[348] **KNIGHT, Thomas Andrew** (1759-1838)
FRS, plant physiologist

Papers (9) submitted to the Royal Society. *Royal Soc*
Papers (2) on birds 1833, 1835. *Linnean Soc*
Paper on descent of sap through the bark, and copies of correspondence (89 items) with Sir J Banks 1795-1819. *Brit Mus (Nat Hist) Botany Dept*
Letters (15) to the Marquess of Bristol 1812-34. *Suffolk RO, Bury St Edmunds* (in Acc 941/56/25, 62, 89) [NRA 6892]
Letters (10) to Sir J Banks 1798-1812. *Brit Libr* (in Add MSS 33980-82)

[349] **LABY, Thomas Howell** (1880-1946)
FRS, physicist

Correspondence, incl some letters from Lord Rutherford. *Private*
Letters (10) to Lord Rutherford 1908-18. *Cambridge Univ Libr* (in MS Add 7653)

[350] **LACK, David Lambert** (1910-1973)
FRS, ornithologist

Papers and correspondence, incl ornithological journals, observations and notes 1925-73, radar observations (6 vols) of bird migration 1957-61, drafts of articles and transcripts of broadcasts. *Edward Grey Inst, Zoology Dept, Oxford* [NRA 18780]
Journal of expeditions to Bear Island, east Greenland and Iceland 1932-33. *Scott Polar Research Inst, Cambridge* (MS 447) [NRA 6527]

See also Alexander, Meinertzhagen, Moreau

[351] **LAMB, Sir Horace** (1849-1934)
FRS, mathematician

Notes for lectures (20 vols) 1896-1920. *John Rylands Univ Libr, Manchester* [NRA 9194]
Letters (16) to Lord Rayleigh 1887-1915. *Private* [NRA 18694]
Letters (14) to Sir GG Stokes 1875-1900. *Cambridge Univ Libr* (in MS Add 7656)
Correspondence (12 items) with Lord Kelvin 1892-1905. *Ibid* (in MS Add 7342)

[352] **LAMBERT, Aylmer Bourke** (1761-1842)
FRS, botanist

Commonplace book containing extracts from natural history publications, drawings, and a meteorological journal, kept at Salisbury *c*1785-1787; catalogues and lists of plants 1803-06; letters addressed to him 1788-1810. *Brit Libr* (Add MSS 28545, 28610-12)
Catalogue of his botanical books 1807; correspondence (2 vols) 1828-40; letters (20) to Sir WJ Hooker 1832-41. *R Botanic Gardens, Kew*
MS of *Genus pinus* (1803-37). *Radcliffe Science Libr, Oxford* (ZB 19) [NRA 12664]
Extracts from botanical papers of P Collinson, tracings of botanical drawings by C Plumier, correspondence (83 items) with Sir JE Smith 1789-1827, and letters to R Pulteney 1796-1800. *Linnean Soc*
Letters (53) to W Cunnington *c*1798-1810. *Wiltshire Archaeological and Nat Hist Soc, Devizes*
Correspondence (18 items) with Harriett Pigott 1828-40. *Bodleian Libr, Oxford* (in MS Pigott d 9)

See also Smith JE

[353] **LANCHESTER, Frederick William** (1868-1946)
FRS, aeronautical and automobile engineer

Papers, photographs and offprints (4 boxes). *Southampton Univ Libr*
Personal papers (*c*800 items). *Lanchester Polytechnic, Coventry*
Papers and correspondence 1916-29, incl a treatise on the part played by skin friction in aeronautics, correspondence with the Air Ministry and War Office 1916, two poems and an autobiography. *Birmingham City Mus* [NRA 9185]
Papers, incl 14 aeronautical reports, and correspondence 1914-39. *R Aeronautical Soc*

[354] **LANG, William Dickson** (1878-1966)
FRS, geologist and entomologist

Correspondence (18 boxes) and papers. *Brit Mus (Nat Hist) Palaeontology Dept*

Diaries 1893, 1943-45, and notes and drawings (2 vols) rel to Charmouth beach. *Dorset RO* (Pavey Colln)
Geological and biological field notes 1920-30, notes on the basking shark 1956, and paper on 'The occurrences of *chirocephalus diaphanus* Prevost on the Devon-Dorset border' 1935. *Dorset County Mus*

[355] **LANKESTER, Sir Edwin Ray,** KCB (1847-1929)
FRS, zoologist

Papers rel to his directorship of the museum 1898-1907; letters (80) to CE Fagan 1898-1920. *Brit Mus (Nat Hist) General Libr*
Unpublished work on evolution theory; letters (13) to K Pearson 1887-1921; student notes on his zoology lectures 1874-91. *University Coll London*
Letters (4 vols) to JR Moir 1910-28. *Brit Libr* (Add MSS 44968-71)
Correspondence (85 items) with Macmillan & Co 1869-1919. *Ibid* (in Add MS 55219)
Correspondence (56 items) with TH Huxley 1872-94. *Imperial Coll, London*
Letters (33) to Sir EW Gosse 1898-1927, and 29 to E Clodd 1899-1918. *Brotherton Libr, Leeds Univ*
Correspondence (20ff) rel to the termination of his lectureship at New College, London 1872. *Dr Williams's Libr, London*
Some correspondence. *Zoology Dept Libr, Oxford*

See also Bateson

[356] **LAPWORTH, Charles** (1842-1920)
FRS, geologist

Correspondence and papers, incl field notebooks, lecture notes and maps. *Geology Dept, Birmingham Univ*
Maps and sections rel to the geology and water supply of the Birmingham area 1892-94. *Birmingham Reference Libr*
Notes on the water supply for the districts of Coalville, Eaton Park and Rutland Derby 1899. *Birmingham Univ Libr* (MS 1959/1) [NRA 9498]
Notes on the Museum of Natural History, Warwick. *Warwick County RO*
MSS and original drawings for his published works on the graptolites of co Down. *Geology Dept, Queen's Univ, Belfast*
Letters from him 1884-1915, incl 13 to FL Kitchin 1913-15. *Inst of Geological Sciences*

[357] **LARMOR, Sir Joseph** (1857-1942)
FRS, physicist

Correspondence (7 vols). *Royal Soc*
Correspondence (*c*510 items) and papers, incl MSS of his published and unpublished works and materials for a work on the dissipation of energy. *St John's Coll, Cambridge* [NRA 22852]
Working papers on electrostatics, and *c*230 items of correspondence 1880-1937. *Cambridge Univ Libr* (in MS 7656)
Correspondence (296 items) with Sir OJ Lodge 1815-1936. *University Coll London*
Correspondence (39 items) with Lord Kelvin 1893-1908. *Cambridge Univ Libr* (in MS 7342)
Correspondence (34 items) with Lord Rutherford 1903-19. *Ibid* (in MS 7653)
Letters (24) to K Pearson 1888-1911. *University Coll London*
Letters (17) to Lord Rayleigh 1896-1917. *Private*
Letters (11) to Sir D Gill 1908-13. *R Geographical Soc*

See also Bateson, Eddington, Fleming JA, Rutherford, Sampson, Schuster, Whittaker

[358] **LAWES, Sir John Bennet,** 1st Bt (1814-1900)
FRS, agriculturist

Papers and correspondence rel to Rothamsted Experimental Station, incl the Lawes Superphosphate Patent Case 1852. *Rothamsted Experimental Station, Harpenden*
Personal papers 1834-94. *Hertfordshire RO* [NRA 10873]
Records of Lawes Chemical Manure Co Ltd from 1872. *Valence Libr, Dagenham*

[359] **LEE, Robert** (1793-1877)
FRS, physician

Diaries (6) 1821-73, with 20 letters to him 1840-66; miscellaneous autobiographical papers. *Wellcome Hist Medical Libr* (MSS 3213-19)
Notes on fibro-calcareous tumours and polypi of the uterus; drawings of specimens illustrating his work on the nerves of the heart and uterus 1828-44. *R Coll of Physicians, London*

[360] **LENNARD-JONES, Sir John Edward,** KBE (1894-1954)
FRS, physicist

Correspondence 1922-54 and papers 1906-54, incl 103 journals 1942-54, notes for and student notes of his theoretical chemistry lectures 1938-52, and other lectures, sermons and addresses delivered by him 1921-54. *Churchill Coll, Cambridge* [NRA 17356]

Correspondence with Viscount Cherwell 1930-46. *Nuffield Coll, Oxford* [NRA 16447]

[361] **LESLIE, Sir John** (1766-1832)
Mathematician and natural philosopher

Mathematical notebooks (3), students' mathematical exercises prescribed by him 1808-19, list of his students 1827, and 122 letters to J Brown 1788-1823. *Edinburgh Univ Libr* (in MSS Dc 1.101, 2.57)
Astronomical notebook, mathematical notebook, and a few items of correspondence 1814-31. *Private* [NRA 9763]
Correspondence mainly with his family and a few papers 1794-1832. *Private* [NRA 21276]
Correspondence (29 items) with JS Langton 1825-29. *Lincolnshire Archives Office* [NRA 8678]

[362] **LHUYD, Edward** (1660-1709)
FRS, naturalist

Papers, incl treatise on materia medica, plant catalogues, lists of shells, coins and medals, translations of Celtic names, notes on chemistry, geology, metals, astronomy, stereometry etc, and extracts from published works. *Brit Libr* (Add MSS 14941, 15053, 15065-77)
Correspondence (10 vols) and papers, incl papers rel to his imprisonment at Brest 1701, proposals for a Celtic dictionary *c*1705, and draft of his sixth lecture 'De stellis marinis' 1706. *Bodleian Libr, Oxford* (MSS Ashmole 1814-16, 1817 a-b, 1820 a-b, 1829-30; MS Eng hist c 11)
Correspondence 1685-1708 and papers, incl list of plants observed in Wales 1682, list of MSS seen at Hengwrt 1696, short Irish-Latin-Welsh vocabulary, and interleaved copy of J Ray's *Synopsis methodica stirpium Britannicarum* (1696) used by him as a travel notebook. *Nat Libr of Wales* (Peniarth MSS 42, 119-20, 251)
Papers (4) submitted to the Royal Society. *Royal Soc*
Printed books annotated and interleaved by him. *Bodleian Libr, Oxford* (Ashm A 15, B 7, C 15; Lister L 92; 8° Rawl 704)
Annotated copy of his *Lithophylacii Britannici ichnographia* (1699). *Brighton Reference Libr*
Plates with MS descriptions from the same. *Brit Mus (Nat Hist) Palaeontology Dept*
Letters, incl 20 to the Revd J Morton 1694-1708, 16 to R Richardson 1697-1705 and 10 to Sir H Sloane 1701-08. *Brit Libr* (Sloane MSS 1039-4076 *passim*)
Letters (102) to M Lister 1689-1702. *Bodleian Libr, Oxford* (MSS Lister 3, 35-37)
Letters (13) to J Aubrey 1693-94. *Ibid* (MS Aubrey 12)
Letters (12) from H Usticke 1700-08. *Ibid* (MS Corn c 1)
▷*Bodleian Library Record*, x, pt 2, 1979, pp112-27
See also Lister M, Ray, Richardson R, Woodward J

[363] **LIGHTFOOT, Revd John** (1735-1788)
FRS, naturalist

Notes on fungi, directions for a tour of Worcestershire, Herefordshire and Gloucestershire, and 16 letters to the Duchess of Portland 1769-72. *Nottingham Univ Libr* (in PwE) [NRA 7628]
Papers (2) submitted to the Royal Society. *Royal Soc*
Annotated copy of his *Flora Scotica* (1777). *Nat Libr of Scotland* (Adv MS 23.5.11)
Annotated copies of J Ray's *Synopsis methodica stirpium Britannicarum*, 3rd edn (1724) and W Hudson's *Flora Anglica* (1762). *Botany Dept, Oxford*
Correspondence with R Pulteney. *Linnean Soc*

LILFORD, Baron, see Powys

[364] **LINDEMANN, Frederick Alexander,** 1st Viscount Cherwell (1886-1957)
FRS, physicist

Official, scientific, political and personal papers and correspondence 1895-1957, incl patents and inventions 1916-57, Clarendon Laboratory records 1919-57, scientific working papers and MSS of his published works, drafts of lectures and speeches, correspondence with foreign scientists and with Sir WS Churchill 1923-57. *Nuffield Coll, Oxford* [NRA 16447]
See also Adam, Adrian, Appleton, Barcroft, Berkeley RTMR, Bragg WL, Chadwick, Cockcroft, Eddington, Egerton, Florey, Fowler, Hume-Rothery, Huxley JS, Jackson H, Jeans, Lennard-Jones, Lodge, Robertson R, Robinson R, Rutherford, Simon FE, Taylor GI, Thomson GP, Tizard, Townsend, Wallis BN, Whittaker

[365] **LINDLEY, John** (1799-1865)
FRS, botanist

Descriptions of plants (9 boxes), annotated copies of his *Rosarum monographia* (1820) and AL de Jussieu's *Genera plantarum secundum ordines naturales disposita* (1789), and botanical drawings. *R Horticultural Soc*
Correspondence and papers, incl report on the royal gardens at Kew, Windsor, Hampton Court and Kensington 1838, botanical correspondence (2 vols) *c*1831-41, official correspondence 1832-54, correspondence rel to FR Chesney's expedition to

Mesopotamia 1835-57; letters (231) to Sir WJ Hooker 1820-59. *R Botanic Gardens, Kew*

Draft of his *Introduction to the . . . natural system of botany* (1830), 74 plant drawings, and correspondence (18 items) with W Hutton 1829-32. *Brit Mus (Nat Hist) Botany Dept*

Correspondence, papers and drawings, with W Hutton, rel to their *Fossil flora of Great Britain* (1831-37). *Hancock Mus, Newcastle upon Tyne*

Annotated proofs (9) of plates from his *Fossil flora of Great Britain*, I (1831). *Inst of Geological Sciences*

Notes on the quercus. *Botany School, Cambridge*

Student notes of his lectures 1830; letters (41) to the Society for the Diffusion of Useful Knowledge 1829-38. *University Coll London* (MS Add 50; SDUK Papers)

Correspondence (20ff) with Sir R Peel 1845-49. *Brit Libr* (Add MSS 40576-40602 *passim*)

Letters (16) to Sir WC Trevelyan *c*1835-65. *Newcastle Univ Libr*

Letters (11) to Sir JFW Herschel 1838-44. *Royal Soc*

[366] **LINSTEAD, Sir Reginald Patrick** (1902-1966)
FRS, chemist

Correspondence 1946-66 and papers, incl notebooks and working papers 1920-55, lectures 1940-63, committee and consultancy papers 1938-60, and administrative papers as rector of Imperial College (1955-66). *Imperial Coll, London* [NRA 19300]

Papers rel to his service as foreign secretary of the Royal Society (1960-65). *Royal Soc*

[367] **LISTER, Joseph,** 1st Baron Lister, OM (1827-1912)
FRS, surgeon

Correspondence and papers, incl 4 commonplace books, lecture notes 1855-65, paper on compound fracture and other writings, notes and drawings on suppuration 1851-1907, and drawings of fungi 1872-77. *R Coll of Surgeons, London* [NRA 9521]

Travel diary 1894, sketch book 1831-34, fragment of an essay on osteology 1843, essay on atomic theory *c*1849, and student notes of his lectures 1871-77. *Wellcome Hist Medical Libr* (MSS 3298-3303, 5018)

Notebooks. *Commonwealth Mycological Inst, Kew*

Clinical case notes (1 vol). *Surgery Dept, Edinburgh Univ*

'On the mode in which external applications act on internal parts' 1855. *R Medical Soc, Edinburgh* (Dissertation Colln)

Notes taken by him of W Sharpey's lectures 1849-50. *Medical Soc of London* (MS 80)

Catalogue of plants collected in Essex 1844-48. *Passmore Edwards Mus, London*

Miscellaneous correspondence and papers 1870-1911, incl *c*140 letters to Sir H Cameron 1879-1911. *R Coll of Surgeons, Edinburgh* [NRA 20261]

Letters (76ff) to Sir HW Acland 1878-99. *Bodleian Libr, Oxford* [NRA 22893]

Letters (27) to A Wilson 1878-1907, and 8 to Sir W Turner 1900-07. *Edinburgh Univ Libr* (in MS Dc 2.96)

Letters (26) 1869-1907. *R Coll of Physicians and Surgeons, Glasgow* [NRA 22540]

Letters to Sir T Barlow, and 14 other letters from him 1897-1909. *R Coll of Physicians, London* [NRA 9520]

Letters (10) to Sir AB Kempe 1894-1907. *West Sussex RO, Chichester* [NRA 17595]

Some correspondence and papers. *Suffolk RO, Ipswich*

[368] **LISTER, Joseph Jackson** (1857-1927)
FRS, zoologist

Drafts and notes for his work on polystomella crispa; laboratory notebooks (6) mainly on foraminifera 1892-1908. *Brit Mus (Nat Hist) Zoology Dept*

Letters (11) to Sir J Burdon-Sanderson 1891-1901. *University Coll London* (in MS Add 179)

[369] **LISTER, Martin** (1639-1712)
FRS, zoologist

Correspondence and papers, incl treatises and notebooks 1660-1710, French travel notebooks 1663-66, 1698, and correspondence (7 vols) 1665-1710. *Bodleian Libr, Oxford* (MSS Lister 1-10, 12-15, 19-40)

Treatise in Latin on British beetles (23ff). *Brit Libr* (Sloane MS 783A)

Papers (8) submitted to the Royal Society; letters (58) to H Oldenburg 1670-77. *Royal Soc*

MS of his commentary on S Sanctorius, *De statica medicina* (1701). *R Coll of Physicians, London*

Annotated copy of his *De cochleis* (1685); notes later incorporated in the third edition of his *Historia conchyliorum* (1770). *Linnean Soc*

Notes later incorporated in the third edition of his *Historia conchyliorum* (1770). *Radcliffe Science Libr, Oxford*

Original drawings (64) for his *Historia conchyliorum*, I (1685). *Science Mus Libr* [NRA 9524]

Letters (88) to E Lhuyd 1690-1709. *Bodleian Libr, Oxford* (MSS Ashm 1816, 1829-30)

See also Grew, Lhuyd, Oldenburg, Ray

[370] **LITTLEWOOD, John Edensor** (1885-1977)
FRS, mathematician

Papers, incl notebooks and files of GH Hardy. *Trinity Coll, Cambridge*
Papers written with GH Hardy on Waring's Problem. *Cambridge Univ Libr* (MS Add 6983)
Correspondence (9 items) with Bertrand Russell *c*1919-1959. *McMaster Univ Libr, Hamilton, Canada*

[371] **LIVERSIDGE, Archibald** (1847-1927)
FRS, chemist

Papers, incl letter books rel to Sydney University science departments, records of the Australian Association for the Advancement of Science, memoranda and photographs. *Sydney Univ Archives*
Illustrated notes taken by him of geology lectures by Sir AC Ramsay, London 1869. *Royal Soc of New South Wales, Sydney*
Correspondence with the Revd WB Clarke. *Mitchell Libr, Sydney*

[372] **LLOYD, Revd Humphrey** (1800-1881)
FRS, man of science

Papers (15 vols), incl papers on magnetism, meteorology and the affairs of Trinity College, Dublin, and correspondence (11 items) 1837-39. *Trinity Coll, Dublin* [NRA 19217]
Correspondence (286 items) with Sir E Sabine and others on terrestrial magnetism; correspondence (51 items) with Sir JFW Herschel 1838-65. *Royal Soc*
Correspondence (114 items) with Sir WR Hamilton 1832-63. *Trinity Coll, Dublin* [NRA 20076]
Correspondence (30 items) with JD Forbes 1833-60. *St Andrews Univ Libr* [NRA 13132]
Letters (15) to Earl Cairns 1868-80. *Public Record Office* (in PRO 30/51/16) [NRA 10007]
Letters (11) to Sir GG Stokes 1849-75. *Cambridge Univ Libr* (in MS Add 7656)

[373] **LLOYD, John Augustus** (1800-1854)
FRS, engineer and surveyor

Account of the Ballane Indians, notes on a map of Darien, and itinerary from Ocaña to Bogotá 1841; comments on plans for a Panama canal 1851; notes on illness and domestic habits of the Pascanians supplementing a report on the silver mines of Cerro de Pasco, 1853. *R Geographical Soc*
'Operations connected with the levellings carried across the Isthmus of Panama'; letters (6) to Sir JFW Herschel 1835-37. *Royal Soc*

[374] **LOCKYER, Sir Joseph Norman,** KCB (1836-1920)
FRS, astronomer

Correspondence and papers, incl laboratory notebooks, mainly spectroscopic, *c*1870-1900, scrapbooks about ballooning 1907-09, solar observations, and spectroscopic plates. *Exeter Univ Libr* (Norman Lockyer Observatory Colln) [NRA 11128]
Correspondence as secretary of the Royal Astronomical Society's Solar Eclipse Committee *c*1870; drawings (16) of planets *c*1852-99; letters, incl 46 to officers of the Society 1863-98 and 19 to its Joint Permanent Eclipse Committee 1898-1911. *R Astronomical Soc*
Correspondence (64 items) with TH Huxley *c*1863-94. *Imperial Coll, London*
Letters (27) to Sir D Gill 1910-12. *R Geographical Soc*
Correspondence (24 items) with Macmillan & Co 1906-19. *Brit Libr* (in Add MS 55218)

See also Frankland, Sylvester

[375] **LODGE, Sir Oliver Joseph** (1851-1940)
FRS, physicist

Correspondence and papers, incl diaries 1871-1940, MSS of his published works, correspondence and papers (440 items) rel to university affairs 1883-1933, and personal and scientific correspondence (2,159 items) 1861-1940. *Birmingham Univ Libr* [NRA 16740]
Correspondence (2,710 items) rel to psychical research 1884-1940. *Incorporated Soc for Psychical Research* [NRA 11857]
Correspondence (1,991 items), mainly scientific, 1871-1938. *University Coll London* (MS Add 89)
Scientific notebooks, papers and letters addressed to him 1874-1912. *Liverpool Univ Libr*
Correspondence and papers (470ff) *c*1883-1930, incl reports, specifications, patents, blueprints, plans, and papers rel to joint litigation with the Marconi Company against the Admiralty. *Inst of Electrical Engineers* [NRA 20573]
Laboratory notebook 1880. *Science Mus Libr*
Letters (94) to Macmillan & Co 1888-1934. *Brit Libr* (in Add MS 55220)
Correspondence (30 items) mainly with the Society of Authors 1909-16. *Ibid* (in Add MS 56739)
Correspondence (49ff) with Gilbert Murray 1904-19. *Bodleian Libr, Oxford* [NRA 16865]
Correspondence (34 items) with Lord Kelvin 1884-1906. *Cambridge Univ Libr* (in MS Add 7342)
Letters (19) to Sir JJ Thomson. *Ibid* (in MS Add 7654)

Letters (10) to Lord Rutherford. *Ibid* (in MS Add 7653)
Letters (20) to HE Armstrong 1885-1935. *Imperial Coll, London* [NRA 11420]
Letters (11) to SP Thompson 1897-1916. *Ibid* [NRA 11421]
Letters (12) to EWW Carlier 1904-19. *Birmingham Univ Libr* [NRA 14359]
Correspondence with Viscount Cherwell 1919-32. *Nuffield Coll, Oxford* [NRA 16447]
Letters (31) to Lord Rayleigh 1882-1908. *Private*
Correspondence with the Earl of Balfour *c*1920-25. *Private* [NRA 10026]

See also Barrett, Crookes, Fitzgerald, Glazebrook, Larmor, Ramsay W, Schuster, Smithells, Strutt JW, Strutt RJ, Thompson SP, Thomson W

[376] **LOGAN, Sir William Edmond** (1798-1875)
FRS, geologist

Geological notebooks (14) and other papers mainly rel to S Wales 1830-42. *Inst of Geological Sciences* (GSMI/67, 218) [NRA 18675]
Travel journals, Italy 1829, France and Spain 1834, Canada and USA 1840-41; journal of the survey of Lake Superior 1846; scrapbook 1845-57. *Toronto Public Libr*
Reports on the geological survey of the Great Lakes 1844-45. *Public Archives of Canada, Ottawa*
Report on Canadian mining locations 1847; history of the geological survey of Canada 1850; correspondence 1837-71. *McGill Univ Libr, Montreal*
Correspondence and papers 1843-70. *McCord Mus, Montreal*
Letters (21) to Sir HT De la Beche. *Nat Mus of Wales Geology Dept, Cardiff*

[377] **LONSDALE, Dame Kathleen,** DBE (1903-1971)
FRS, crystallographer

Correspondence and scientific papers, incl MSS of her published works, papers rel to the Society of Friends etc. *Private*
Correspondence (1 file) with WT Astbury 1942-53. *Brotherton Libr, Leeds Univ* [NRA 19429]
Correspondence with CA Coulson 1949-60. *Bodleian Libr, Oxford* [NRA 21828]
Correspondence with H Dingle rel to the Royal Society 1958-*c*1971. *Imperial Coll, London* [NRA 24527]

[378] **LOW, Revd George** (1746-1795)
Naturalist

MSS of his published and unpublished works on the fauna, flora and history (3 vols) of Orkney, 'Miscellanea No I. A cabinet of curiosities collected by GL' 1766-68, illustrated natural history essays 1770, a course of microscopical observations 1771-72, and 'A tour thro' the islands of Orkney and Shetland' 1774. *Edinburgh Univ Libr* (MSS Dc 6. 102-07; MS La III 580)
Papers, including MS of his *Fauna Orcadensis* (1813), 'A chronological sketch of the history of the Orkneys', 'A description of Zetland', 'Some observations on natural history' 1770, lists of plants, animals and fossils, drawings of plants; letters (56) to G Paton 1772-95. *Nat Libr of Scotland* (Adv MSS 29.5.8, 31.3.1, 32.4.1; MS 3935)
'A tour in the islands of Orkney and Schetland, 1774 and 1778', 'History and description of Orkney, and also a tour of the Fair Isle and Shetland' *c*1780. *R Scottish Mus, Edinburgh*
Letters (20) to T Pennant 1772-75. *Warwick County RO* (TP 290/1-20) [NRA 23685]

[379] **LOWE, Revd Richard Thomas** (1802-1874)
Naturalist

'Fungi maderenses' (1 vol); botanical lists rel to Madeira and the Canary Islands; letters (1 vol) addressed to him 1828-70; letters (62) to Sir WJ Hooker 1827-65. *R Botanic Gardens, Kew*
Papers on fishes and invertebrates of Madeira 1850-61. *Brit Mus (Nat Hist) Zoology Dept*
Flora of Madeira. *Botany School, Cambridge*

[380] **LUBBOCK, John,** 1st Baron Avebury (1834-1913)
FRS, entomologist

Correspondence and papers (44 vols) 1855-1913. *Brit Libr* (Add MSS 49638-81)
Notebooks (25). *Royal Soc*
Diary. *Private*
Letters (2 vols) to Macmillan & Co 1873-1913. *Brit Libr* (Add MSS 55213-14)
Letters (45) to Sir GG Stokes 1873-1901. *Cambridge Univ Libr* (in MS Add 7656)
Letters (23) to TH Huxley 1856-95. *Imperial Coll, London*
Correspondence (17 items) with Sir E Chadwick 1864-89. *University Coll London* [NRA 21653]

[381] **LUBBOCK, Sir John William,** 3rd Bt (1803-65)
FRS, astronomer and mathematician

General correspondence (42 vols); correspondence (63 items) with Sir JFW Herschel. *Royal Soc*

Letters (over 100) to the Society for the Diffusion of Useful Knowledge 1829-43. *University Coll London*

Letters (47) to C Babbage 1830-61. *Brit Libr* (in Add MSS 37185-37200)

Correspondence (29 items) with Sir WR Hamilton 1832-53. *Trinity Coll, Dublin* (in MSS 1493, 5123-33) [NRA 20076]

Letters (23) to officers of the Royal Astronomical Society 1830-63. *R Astronomical Soc*

See also Adams, Airy, Babbage, Baily, Barlow, Beaufort, Brande, Brodie BC the elder, Daniell, Darwin CR, De Morgan, Fitton, Gassiot, Gilbert, Hamilton WR, Ivory, Lyell, Murchison, Peacock, Powell B, Sabine, Sheepshanks, Somerville, Stratford, Talbot, Whewell

[382] **LYDIAT, Revd Thomas** (1572-1646)
Chronologer

Correspondence and papers, incl mathematical notes and treatises (2 vols), MSS of his published and unpublished works on chronology (8 vols), Gospel harmonies in Hebrew, Greek and English, and 63 items of correspondence 1603-44. *Bodleian Libr, Oxford* (MSS Bodl 313, 658, 661-64, 666-71; SC 27558, 27565)

Unpublished theological treatises (2), and MS of his published treatise on the Arundel marbles. *Trinity Coll, Dublin*

[383] **LYELL, Sir Charles,** 1st Bt (1797-1875)
FRS, geologist

Correspondence and papers (13 boxes), incl notes on the geology of Angus, Madeira, the Canary Islands, New Zealand and Mount Etna 1825-58, notes on sea serpents 1845-48 and for lectures and speeches 1833-64, and *c*2,220 items of correspondence 1824-74. *Edinburgh Univ Libr* (MSS Gen 108-20, MSS Lyell 1-10) [NRA 9505]

Correspondence and papers, incl 7 scientific journals 1855-61 and other notebooks. *Private*

Letters (*c*450) from CR Darwin 1837-74, with *c*240 others addressed to him *c*1808-1874. *American Philosophical Soc, Philadelphia*

Letters (2 vols) to GA Mantell 1821-51. *Alexander Turnbull Libr, Wellington*

Letters (136) to C Babbage 1826-54. *Brit Libr* (Add MSS 37183-37200 *passim*)

Letters (135) to Sir WJ Hooker 1828-45. *R Botanic Gardens, Kew*

Correspondence (*c*80 items), incl 25 letters to Sir WR Grove and *c*20 to J Tyndall. *Royal Inst*

Correspondence (75 items) with TH Huxley 1853-73. *Imperial Coll, London*

Correspondence (52 items) with Sir JFW Herschel 1825-68; letters (12) to Sir JW Lubbock 1835-56. *Royal Soc*

Letters (50) to Sir RI Murchison. *Geological Soc of London*

Correspondence (41 items) with JD Forbes 1832-65. *St Andrews Univ Libr* [NRA 13132]

Letters (31ff) to W Whewell 1831-63. *Trinity Coll, Cambridge* [NRA 8804]

Correspondence with John Murray (Publishers) Ltd. *Private*

Letters (21) to Sir HT De la Beche. *Nat Mus of Wales Geology Dept, Cardiff*

Letters (11) rel to the publication of his *Manual of elementary geology* 1854-57. *Private*

Letters (10) to Mary Somerville 1831-*c*1869. *Bodleian Libr, Oxford* [NRA 9423]

▷LG Wilson, *Charles Lyell. The years to 1841: the revolution in geology*, 1972

See also Bates, Carpenter WB, Darwin CR, Davidson, Falconer, Forbes E, Godwin-Austen, Herschel JFW, Huxley TH, Ramsay AC, Woodward SP

[384] **LYONS, Israel** (1739-1775)
Mathematician, astronomer and botanist

MS of his *Treatise of fluxions* (1758) and other published and unpublished works on mathematics and astronomy *c*1756-75. *R Greenwich Observatory, Herstmonceux* (MSS 292-99) [NRA 22822]

Botanical observations, with M Tyson, in the neighbourhood of Cambridge 1760-70. *Suffolk RO, Bury St Edmunds*

Interleaved and annotated copy of W Hudson's *Flora Anglica* (1762). *Linnean Soc*

[385] **McCOY, Sir Frederick,** KCMG (1817-1899)
FRS, palaeontologist

Papers and correspondence as director of the National Museum 1858-99. *Nat Mus of Victoria, Melbourne*

Scientific correspondence (1 vol) 1841-92, with testimonials and biographical material. *Mitchell Libr, Sydney*

Letters (11) from Sir FJH von Müller 1859-89. *R Botanic Gardens and National Herbarium, South Yarra, Australia*

See also Gould

[386] MacGILLIVRAY, John (1822-1867)
Naturalist

MS of his *Narrative of the voyage of HMS Rattlesnake . . . 1846-50* (1852); catalogue of botanical specimens collected during the voyage; letters (11) to Sir WJ Hooker 1848-54. *R Botanic Gardens, Kew*
Journal (2 vols) of voyage of HMS *Herald* 1852-55; vocabularies of Pacific islands. *Naval Hist Libr, London* (MSS 23, 73/4, 74/3)
Notebooks containing notes on the natural history of Cape York and northern Australia and copies of letters written on board HMS *Rattlesnake* 1847-49. *Private*
Catalogue of radiata and mollusca collected during the voyage of HMS *Rattlesnake* 1846-50. *Brit Mus (Nat Hist) Zoology Dept*

[387] MacGILLIVRAY, William (1796-1852)
Naturalist

Hebridean journal 1817-18, travel notes 1819, diary 1833, notes mainly on British birds 1836-40, and student notes of his natural history lectures 1840-45. *Aberdeen Univ Libr*
Drawings (213) of British mammals, birds and fishes 1831-41. *Brit Mus (Nat Hist) Zoology Dept*
Correspondence with T Edmondston 1840-46. *Shetland Libr, Lerwick* [NRA 13501]

[388] McINTOSH, William Carmichael (1838-1931)
FRS, zoologist

Correspondence (*c*1,500 items) and papers, incl lecture notes, exercise books, botanical specimens, diaries, and accounts 1852-1931. *St Andrews Univ Libr* (MSS 37097-37120)
Notes and sketches rel to annelids. *Brit Mus (Nat Hist) Zoology Dept*

[389] McKIE, John (*c*1820-1915)
Engineer

Memoirs (5 vols). *Nat Libr of Scotland* (Acc 3420)

[390] McLACHLAN, Robert (1837-1904)
FRS, entomologist

Entomological correspondence. *Hope Libr, University Mus, Oxford*
Drawings (65) for his *Monographic revision and synopsis of the trichoptera of the European fauna* (1874-80). *Brit Mus (Nat Hist) Entomology Dept*
Miscellaneous correspondence and certificates. *R Entomological Soc*

[391] MACLAURIN, Colin (1698-1746)
FRS, mathematician

Memorial (72pp) to the Commissioners of Excise on the mensuration of ships' cargoes 1735; MSS of his *Treatise of algebra* and *Treatise of fluxions;* student notes of his lectures. *Edinburgh Univ Libr* (MSS Dc 1.17, 3.66, 7.73; MS Gen 75D)
Memorial to the Commissioners of Excise on the mensuration of ships' cargoes 1735. *Nat Libr of Scotland* (Adv MS 23.1.13)
Papers (5) submitted to the Royal Society, and miscellaneous letters (9) 1720-37. *Royal Soc*
Rule for finding the meridional parts to any spheroid 1741. *Brit Libr* (in Add MS 4437)
'Nugae poeticae'. *Ibid* (Add MS 52247)
Letters (11) to J Stirling 1728-40. *Private* [NRA 14810]

[392] MACLEAR, Sir Thomas (1794-1879)
FRS, astronomer

Letter books, correspondence and papers from 1811. *Cape Archives Depot, Cape Town* (Acc 515)
Correspondence and papers as astronomer royal, Cape of Good Hope (1833-70), incl weekly administrative reports, observations, computations, letter books, correspondence, and some personal papers. *R Greenwich Observatory, Herstmonceux*
Correspondence and papers (49 items), incl astronomical observations 1834-35 and correspondence with J Lee 1833-64. *Buckinghamshire RO, Aylesbury* (D/LE/H/8/1-37) [NRA 10775]
'On the coast surveys of South Africa' 1855. *R Geographical Soc*
Letters (57) from D Livingstone 1852-72. *Nat Archives of Zimbabwe, Salisbury*
Correspondence (259 items) with Sir JFW Herschel 1833-70. *Royal Soc*

[393] McLENNAN, Sir John Cunningham, KBE (1867-1935)
FRS, physicist

Correspondence and papers (7 vols) 1867-1935, incl scrapbooks, certificates and diplomas. *Toronto Univ Libr*
Correspondence (30 items) with N Bohr 1911-24. *Niels Bohr Inst, Copenhagen*
Letters (15) to Lord Rutherford 1902-35. *Cambridge Univ Libr* (in MS 7653)

MAKDOUGALL BRISBANE, see Brisbane

[394] MANTELL, Gideon Algernon (1790-1852)
FRS, geologist

Correspondence and papers 1801-52, incl journal (4 vols) 1819-52, description (6 vols) of European tour 1833-35, medical case notes (4 vols) 1811-51 and other medical notes from 1809, geological notes and sketches, catalogue of his fossil collection 1839, poems and other published and unpublished writings. *Alexander Turnbull Libr, Wellington*

Catalogue of his fossil collection, water-colour drawings of fossils, and letters (1 vol) to RTWL Brickenden on the telerpeton 1848-52. *Brit Mus (Nat Hist) Palaeontology Dept*

Illustrated paper on the strata in the vicinity of Lewes *c*1812. *Geological Soc of London*

Miscellaneous letters (27). *Sussex Archaeological Soc, Lewes*

Correspondence with A Cuvier and A Brouquiart. *Mus National d'Histoire Naturelle, Paris*

Letters (15ff) to the Marquess of Northampton 1832-45. *Private* [NRA 21088]

Correspondence (13 items) with J Hawkins 1820-36. *West Sussex RO, Chichester*

Letters (12) to the Revd W Buckland 1825-47. *Royal Soc*

See also Lyell

[395] **MARSHALL, Arthur Milnes** (1852-1893)
FRS, naturalist

Notebooks 1879-93. *Zoology Dept, Manchester Univ*

[396] **MARSHALL, John** (1818-1891)
FRS, anatomist

Letters addressed to him (4 vols). *Exeter Univ Libr*

Letters (4) to Sir GG Stokes 1869-89. *Cambridge Univ Libr* (in MS Add 7656)

[397] **MARTIN, Sir Charles James** (1866-1955)
FRS, physiologist

Notebooks (3) and letters to Dame Harriette Chick rel to experiments in collaboration with the Lister Institute 1936-38; correspondence with JT Wilson 1895-1929; letters from hospital in Lemnos and Egypt 1915-16. *Australian Academy of Science, Canberra*

Correspondence and papers rel to the Lister Institute, research topics and administration. *Medical Research Council*

[398] **MARTYN, Revd Thomas** (1735-1825)
FRS, botanist

Letters (1 vol) from him and his father J Martyn. *Brit Mus (Nat Hist) Botany Dept* (Banksian MS 103)

Annotated copy of C Linnaeus's *Species plantarum*, 2nd edn (1762-63), and letters incl 24 to Sir JE Smith 1793-1819. *Linnean Soc*

Annotated copy of his *Catalogus horti botanici Cantabrigiensis* (1771). *Botanic Garden, Cambridge*

Letters (26ff) from J Strange 1769-86. *Brit Libr* (Add MS 33349)

Illustrations (88) for his *Universal conchologist* (1784-87). *Brit Mus (Nat Hist) Zoology Dept*

[399] **MASERES, Francis** (1731-1824)
FRS, mathematician

Papers, incl mathematical problems and calculations, translations, notes on France, Russia and Canada, annotations and additions to printed works. *Inner Temple Libr, London*

Logarithms (1 vol). *University Coll London* (MS Graves 30)

Papers (4) submitted to the Royal Society. *Royal Soc*

Mathematical notes; letters (9) from W Morgan 1799-1800. *Edinburgh Univ Libr* (MS Dc 5.7)

Supplement to his 1766 tract on the government of Quebec, 1768. *Univ of British Columbia Libr, Vancouver*

His accounts as cursitor baron of the exchequer 1774-78. *Lewis Walpole Libr, Farmington, Connecticut* [NRA 22338]

Letters (29) to SP Rigaud 1804-16. *Bodleian Libr, Oxford* (in MS Rigaud 61)

[400] **MASKELYNE, Revd Nevil** (1732-1811)
FRS, astronomer

Correspondence and papers 1743-1811, incl astronomical and meteorological observations, computations, tables and catalogues, astronomical and mathematical essays, journals of voyages to St Helena, India and the E Indies, and papers rel to optics, poetry and the *Nautical Almanac*. *R Greenwich Observatory, Herstmonceux* (MSS 136-286, 1234) [NRA 22822]

Correspondence and papers, incl 12 account and memoranda books 1763-1805, papers (24 items) rel to the Royal Observatory 1760-70, and letters (1 folder) addressed to him. *Wiltshire RO, Trowbridge* [NRA 23462]

Observations 1765-1810, miscellaneous correspondence (16 items), and correspondence (10 items) with Sir J Banks 1791-1809. *Royal Soc*

Correspondence (70 items) with Sir W Herschel 1781-1808. *R Astronomical Soc*

[401] MASSON, Sir David Orme, KBE (1858-1937)
FRS, chemist

Correspondence and papers (1 box), incl MSS of his published works, printed testimonials 1884-1916, press cuttings, obituaries, photographs 1893-1933 and other biographical material. *Australian Academy of Science, Canberra*

Paper with M Hay on nitroglycerine 1883. *Wellcome Hist Medical Libr* (MS 2796)

[402] MATTHEY, George (1825-1913)
FRS, chemist

Letter books of Johnson, Matthey & Co Ltd, assayers, refiners and bullion merchants, incl Matthey's correspondence on metallurgical problems 1851-1906. *Private*

Letters (5) to Sir GG Stokes 1889-90. *Cambridge Univ Libr* (in MS Add 7656)

[403] MAXWELL, James Clerk (1831-1879)
FRS, physicist

Correspondence and papers 1847-79, incl laboratory and other notebooks, notes of lectures attended, MSS of his published works, notes and papers on astronomy, colour and optics, electricity and magnetism, geometry, mechanics and dynamics, molecular physics and gases, the properties of matter and general physics. *Cambridge Univ Libr* (MS Add 7655) [NRA 24528]

Scientific notebooks (3) 1860-70, incl draft of his *Treatise on electricity and magnetism* (1873). *King's Coll, London*

'Observations on circumscribed figures' submitted to the Royal Society of Edinburgh 1846. *Royal Soc of Edinburgh*

Miscellaneous papers, and 38 letters to his family 1845-79. *Peterhouse, Cambridge* [NRA 9500]

Letters (41) to Sir GG Stokes 1853-79. *Cambridge Univ Libr* (in MS Add 7656)

Correspondence (33 items) with CJ Monro 1855-71. *Greater London RO* [NRA 17730]

Letters (28) to Lord Kelvin 1857-73. *Glasgow Univ Libr*

Letters (24) to Lord Kelvin 1854-79. *Cambridge Univ Libr* (in MS Add 7342)

See also Fleming JA, Stokes, Tait, Thomson W

[404] MAYERNE, Sir Theodore Turquet de (1573-1655)
Physician

Correspondence and papers 1585-1655, incl notebook on logic kept as a schoolboy at Geneva 1585, pharmacopoeia, formulae for remedies, case notes, medical diary 1611, discourses and collections rel to medicine and chemistry, prescriptions for the British royal family 1612-44, and copies of *Pharmacopoeia Londinensis* (1618 and 1627) annotated by him. *Brit Libr* (777. i. 11; 777. k. 7; Sloane MSS *passim;* Add MS 46378)

Papers 1607-55, incl notebooks and journal. *Cambridge Univ Libr* (MSS Dd iv 21, 25, 33; v 25, 26; ix 85; x 65)

Referee report 1630, report, with D Bethune and M Lister, recommending the erection of pest-houses 1631, petitions (3) 1633-38, certificates (3) 1638-42, and letters (17) 1611-52. *Public Record Office* (SP 14-18 *passim*)

Case book 1607-51. *R Coll of Physicians, London*

'Antidotarium' 1606. *Bodleian Libr, Oxford* (MS Rawl C 51 b)

'Viaticum sive medicorum experimentorum formulae' 1621, with history of his family prefixed. *R Coll of Surgeons, London*

See also Willis T

[405] MEINERTZHAGEN, Richard (1878-1967)
Ornithologist

Diaries (76 vols) kept in India, Mauritius, E Africa and the Middle East 1899-1965; papers rel to his official service 1902-24. *Rhodes House Libr, Oxford*

Catalogue (42 vols) of his bird collection, with ecological notes on birds. *Brit Mus (Nat Hist) Zoology Dept*

Corrected typescript of his *Kenya diary, 1902-1906* (1957). *Nat Libr of Scotland* (Acc 4946)

Correspondence with FM Bailey 1928-60. *India Office Libr and Records* [NRA 21883]

Correspondence with DL Lack c1941-66. *Edward Grey Inst, Zoology Dept, Oxford* [NRA 18780]

Correspondence with RE Moreau 1951-54. *Ibid* [NRA 18632]

Correspondence with HStJ Philby and his wife 1945-52. *Middle East Centre, St Antony's Coll, Oxford*

[406] MELDOLA, Raphael (1849-1915)
FRS, chemist

Correspondence and papers 1864-1908. *Passmore Edwards Mus, London* [NRA 12102]

Notes (7 vols) on lectures attended and given by him 1866-90 and on his research 1879-1906, diary 1875, and miscellaneous correspondence 1899-1907. *Imperial Coll, London* [NRA 22853]

Notes for his English translation of A Weismann's *Studies in the theory of descent* (1880-82). *Hope Libr, University Mus, Oxford*

[407] **MERCER, John** (1791-1866)
FRS, chemist

Autobiography, diplomas, certificates and US patent grant. *Clayton-le-Moors Public Libr*
Letters (17) to Lord Playfair 1842-44. *Manchester Literary and Philosophical Soc*

[408] **MEYRICK, Edward** (1854-1938)
FRS, entomologist

Descriptions, drawings and water-colour paintings of micro-lepidoptera. *Brit Mus (Nat Hist) Entomology Dept*

[409] **MICKLEBOROUGH, Revd John** (*c*1691-1756)
Chemist

Lectures on chemistry. *Gonville and Caius Coll, Cambridge* (MS 619)

[410] **MIERS, Sir Henry Alexander** (1858-1942)
FRS, mineralogist

Correspondence and papers, incl autobiographical notes 1938-41, diaries (3) of visits to Sweden, Russia and S Africa 1892-1903, pocket diaries (17) 1926-42, essay on 'The alchemist' *c*1876, notes on J Chevalier's autobiographical sketch. *Bodleian Libr, Oxford* (MSS Eng misc c 539, d 661, f 415-17, g 79-96)
Papers and letters. *Private*

[411] **MIERS, John** (1789-1879)
FRS, botanist and engineer

Papers, incl 'A catalogue of the woods of Brazil', catalogue of Ceylon plants sent by G Gardner to Sir WJ Hooker 1845, drawings of S American plants and their dissections, and annotated copies of published works. *Brit Mus (Nat Hist) Botany Dept*
Entomological catalogues. *Hope Libr, University Mus, Oxford*
Letters (50) to Sir WJ Hooker 1826-57. *R Botanic Gardens, Kew*

[412] **MILNE, John** (1850-1913)
Mining engineer and seismologist

Notebooks, diagrams and letters on seismology. *Isle of Wight RO, Newport*

[413] **MILNER, Revd Isaac** (1750-1820)
FRS, mathematician

His Jacksonian lectures 1784-87. *Cambridge Univ Libr* (MS Ee v 35)
Papers (4) submitted to the Royal Society. *Royal Soc*
Letters (9) to W Wilberforce *c*1792-1807. *Bodleian Libr, Oxford* (in MSS Wilberforce c 3, d 13, 15, 17) [NRA 7132]

[414] **MIVART, St George Jackson** (1827-1900)
FRS, biologist

Letters (16) to H Allon 1875-84. *Dr Williams's Libr, London* (p24/110) [NRA 13168]
Correspondence (13 items) with TH Huxley 1874-86. *Imperial Coll, London* [NRA 8194]
Letters to ACLG Günther. *Brit Mus (Nat Hist) General Libr*

[415] **MOLYNEUX, William** (1656-1698)
FRS, natural philosopher

Papers, mainly rel to the Dublin Philosophical Society. *Trinity Coll, Dublin* (MSS 888-89) [NRA 19217]
Papers and some correspondence as secretary of the Dublin Philosophical Society 1683-86. *Brit Libr* (Add MS 4811)
Translations of Galileo's third and fourth dialogues; correspondence (66 items) with J Flamsteed 1681-90. *Southampton City RO* (D/M 1/1, 4/15-16) [NRA 4957]
Some papers and correspondence, incl a paper submitted to the Oxford Philosophical Society 1686. *Trinity Coll, Cambridge* (in MS R 4.45)
Papers (6) submitted to the Royal Society, and letters to F Aston 1684-85. *Royal Soc*
Annotated copy of his *Dioptrica nova* (1692). *Brit Libr* (537 k.17)
Letters (34) to J Locke 1692-98. *Carl H Pforzheimer Libr, New York*
Letters (30) from J Locke 1692-98. *Private*

[416] **MONRO, Alexander (I)** (1697-1767)
FRS, physician

Treatises on anatomy, lectures, student notebook and other papers 1717-*c*1753. *Otago Univ Medical Libr, Dunedin*
History of anatomy 1733; 'An account of the operations of surgery' 1748. *R Coll of Surgeons, London*
History of anatomy; autobiography; student notes of his lectures. *Edinburgh Univ Libr*
'An essay on female conduct' *c*1740. *Nat. Libr of Scotland* (MS 6658)

▷RE Wright-St Clair, *Doctors Monro: a medical saga*, 1964

[417] **MONRO, Alexander (II)** (1733-1817)
Anatomist

Papers, incl case notes 1767-1811, papers and lectures on anatomy, physiology etc 1759-1811. *Otago Univ Medical Libr, Dunedin*

'A treatise of wounds and tumours' (2 vols); student notes of his lectures. *R Coll of Surgeons, London*

'A description of the rickets'; student notes of his lectures. *Edinburgh Univ Libr*

▷RE Wright-St Clair, *Doctors Monro: a medical saga*, 1964

See also Hope TC

[418] **MORAY, Sir Robert** (1608-1673)
FRS, natural philosopher

Correspondence with the Marquess of Tweeddale and the Duke of Lauderdale. *Nat Libr of Scotland* (in MSS 7003-06, 7022-25)

Letters (116) to the Earl of Kincardine 1657-73. *Private*

Letters (19) to H Oldenburg 1665-66; papers (19) submitted to the Royal Society. *Royal Soc*

[419] **MOREAU, Reginald Ernest** (1897-1970)
Ornithologist

Correspondence 1930-70 and papers, incl diaries (3) 1918-62, observations (88 vols) of nests in Africa 1937-46 and of quails 1953, and notes, drafts and correspondence rel to his *Palaearctic-African bird migration systems* (1972). *Edward Grey Inst, Zoology Dept, Oxford* [NRA 18632]

Correspondence with DL Lack *c*1943-70. *Ibid* [NRA 18780]

See also Meinertzhagen

[420] **MORISON, Robert** (1620-1683)
Botanist

'Nomenclator stirpium'; annotated copies of A Brunyer's *Hortus regius Blesensis* (1653) and A Caesalpinus' *De plantis libri xvi* (1583). *Bodleian Libr, Oxford* (MSS Sherard 26-28) [NRA 6305]

See also Bobart

[421] **MOSELEY, Henry Gwyn Jeffreys** (1887-1915)
Physicist

Papers rel to his life and work. *History of Science Mus, Oxford* (MS Mus 118)

Family letters (117) 1897-1915. *Private*

Letters (10) to Lord Rutherford 1910-15. *Cambridge Univ Libr* (in MS Add 7653)

▷JL Heilbron, *HGJ Moseley, the life and letters of an English physicist 1887-1915* (1974)

[422] **MURCHISON, Sir Roderick Impey,** 1st Bt (1792-1871)
FRS, geologist

Correspondence and papers, incl field notebooks, diaries and autobiographical journal (162 vols) 1814-65, MSS of his published and unpublished writings and lectures, maps, drawings, and over 1,000 letters addressed to him 1824-71. *Geological Soc of London*

Correspondence and papers, incl 14 field notebooks 1817-68, MSS of his published works, maps, sections, official correspondence and papers as director general of the Geological Survey 1855-71, and some miscellaneous correspondence 1836-71. *Inst of Geological Sciences* [NRA 18675]

Geological and botanical notebooks, cash books etc (14 vols), and *c*80 letters to his wife 1829-44. *Private*

Geological notebook rel to Devon 1836. *Sedgwick Mus of Geology, Cambridge*

Maps and sections of Scottish rocks 1827. *Geology Dept, University Coll London*

Letters addressed to him (1,855ff) 1820-71. *Brit Libr* (Add MSS 46125-28)

Correspondence (*c*200 items) 1827-70, incl 17 letters to his wife 1854-57 and correspondence and papers rel to the discovery of gold in Australia 1850-55. *Edinburgh Univ Libr* (MS Gen 523)

Letters (388) to officers of the Royal Geographical Society 1839-68, with letters to Sir G Back, comments on papers submitted to the Society for publication etc. *R Geographical Soc*

Correspondence with A Sedgwick, 90 letters to G Featherstonhaugh 1827-61 and 50 to R Harkness 1859-61. *Cambridge Univ Libr* (in MS Add 7652)

Correspondence (122 items) with W Whewell 1830-62. *Trinity Coll, Cambridge* [NRA 8804]

Correspondence (103 items) with JD Forbes 1831-67. *St Andrews Univ Libr* [NRA 13132]

Letters (80) to the Revd G Gordon 1832-67. *Private* [NRA 22745]

Letters (68) from D Livingstone 1856-68. *Nat Archives of Zimbabwe, Salisbury*

Correspondence (61 items) with Sir JFW Herschel 1830-68; letters (10) to Sir JW Lubbock 1830-42. *Royal Soc*

Letters (40) to C Babbage 1829-57. *Brit Libr* (Add MSS 37184-37201 *passim*)

Letters (30) from the Revd W Buckland. *Devon RO, Exeter*

Correspondence (25 items) with TH Huxley 1855-69. *Imperial Coll, London*

Letters (22) to Sir HT De la Beche 1833-54. *Nat Mus of Wales Geology Dept, Cardiff*

▷JC Thackray, 'Essential source-material of Roderick Murchison', *J Soc Bibliog Nat Hist*, vi, pt 3, 1972, pp162-70

See also Buckland, Clarke WB, Geikie A, Herschel JFW, Horner, Lyell, Owen, Phillips J, Sabine, Sedgwick, Whewell

[423] **MURRAY, Sir John,** KCB (1841-1914) FRS, naturalist and oceanographer

Records of the voyage of HMS *Challenger* incl his diary 1873-76; correspondence, descriptions of soundings etc rel to other expeditions. *Brit Mus (Nat Hist) Mineralogy Dept*

Progress reports, estimates, and correspondence with the Challenger Publication Committee. *Royal Soc*

Papers and correspondence (1 vol) rel to the voyage of HMS *Challenger. R Botanic Gardens, Kew*

Letters (217) to AE Agassiz 1877-1910, and memorial address on Agassiz 1911. *Edinburgh Univ Libr* (in MS Dk 6.13)

Letters (35) to HR Mill. *Scott Polar Research Inst, Cambridge* (in MS 100) [NRA 6527]

Letters to ACLG Günther 1883-1911. *Brit Mus (Nat Hist) General Libr*

See also Andrews CW

[424] **MUSGRAVE, William** (?1655-1721) FRS, physician

Papers and correspondence rel to the Oxford Philosophical Society, incl his journal of its proceedings 1683-90. *Bodleian Libr, Oxford* (MSS Ashmole 1810-13)

Letters addressed to him rel to natural history 1685-86. *Trinity Coll, Cambridge*

Paper submitted to the Royal Society; letters (12) 1686-1714. *Royal Soc*

Annotated copy of *Pharmacopoeia Bateana* (1691). *Queen's Coll, Oxford* (MS 64)

Letters (28) to Sir H Sloane 1698-1718. *Brit Libr* (in Sloane MSS 4025, 4037, 4039-45, 4060)

[425] **NAPIER, John** (1550-1617) Mathematician

List of projected defensive weapons, incl burning mirrors to destroy enemy ships, heavy artillery and submarine craft, 1596. *Lambeth Palace Libr* (in MS 658)

See also Foster S

[426] **NASMYTH, James** (1808-1890) Engineer

Business records, incl correspondence and working drawings, of Nasmyth, Gaskell & Co, engineers, from 1836. *Eccles Central Libr*

Autobiography (100pp). *Private*

Drawings (2 vols); letters (119) addressed to him. *Edinburgh Central Libr*

Sketch book; correspondence (8 items). *Inst of Mechanical Engineers*

Diagram of marine battering ram. *Nat Maritime Mus, Greenwich*

Letters (17) to H Bicknell 1859-79. *Nat Libr of Scotland* (Acc 4953-54)

Letters (13) to Sir JFW Herschel 1845-71. *Royal Soc*

[427] **NEWTON, Alfred** (1829-1907) FRS, zoologist

Correspondence and papers, incl notes with J Wolley on the great auk and with E Newton on weather and birds at Elvedon, Suffolk, 1850-60; correspondence and papers rel to the great bustard and to the Wild Birds Protection and Preservation Acts. *Balfour and Newton Libr, Zoology Dept, Cambridge*

Correspondence and papers, incl catalogues of birds, bird bones and eggs in Cambridge University Museum of Zoology, notes (2 vols) on birds' eggs from 1855, notes and drawings for his lectures and published works, diplomas, correspondence (4 vols) with JH Gurney and JH Gurney jr 1853-90, and letters from Catherine Strickland 1863-88. *Mus of Zoology, Cambridge*

Lectures on zoology 1866-75. *Cambridge Univ Libr* (MS Add 6415)

Letters (*c*1,000) from JH Gurney and his son 1853-1907. *Norfolk RO*

Some correspondence, incl letters to ACLG Günther 1859-1906. *Brit Mus (Nat Hist) General Libr*

Letters (2 bundles) to RH White 1898-1907. *Gilbert White Mus, Selborne* (White MSS 1143-44)

Letters (26) to Sir PHM Bahr 1903-07. *Edward Grey Inst, Zoology Dept, Oxford*

Letters (19) to H Scherren 1904-06. *Brit Libr* (in Add MS 38794)

Letters (13) to Sir WJ Hooker 1860-65. *R Botanic Gardens, Kew*

See also Darwin CR, Gould, Salvin

[428] **NEWTON, Sir Isaac** (1642-1727) FRS, natural philosopher

[Newton's papers passed by marriage to the Wallop family, Earls of Portsmouth, who presented a large part to Cambridge University Library in 1872; see *Catalogue of the Portsmouth Collection*, Cambridge, 1888. The remainder, described in HMC *8th Report*, App I, 1881, pp60-92, were dispersed by auction in 329 lots at Sotheby's, 13 July 1936]

Correspondence and papers 1661-1725, incl commonplace books, lectures, papers on mathematics, optics, astronomy, mechanics, hydrostatics and chemistry, MSS and annotated copies of the *Principia* and related papers. *Cambridge Univ Libr*

Correspondence (80 items) and papers, incl essays and notes on religious topics, monarchies, alchemy, and on King's College. *King's Coll, Cambridge* (Keynes Newton MSS 1-152)
Papers and correspondence rel to theology, chemistry and alchemy. *Jewish Nat and Univ Libr, Jerusalem* (Yahuda Colln)
Papers (5 vols) as Warden and Master of the Mint, incl papers on science, history and theology, *c*1696-1727. *Public Record Office* (Mint 19)
Papers on chronology, biblical criticism, and church history. *Bodleian Libr, Oxford* (New Coll MS 361)
Correspondence and papers, incl papers for and correspondence rel to *Commercium epistolicum* (1712) and 14 letters to H Oldenburg 1671-76. *Royal Soc*
Miscellaneous correspondence and papers, incl accounts (1 vol), correspondence (74 items) with R Cotes 1709-15 and copies (51) of his and other published works. *Trinity Coll, Cambridge*
Correspondence and papers, incl essays and notes rel to chemistry, alchemy, religion, parliamentary procedure and genealogy. *Babson Coll, Babson Park, Mass*
Correspondence and papers. *Smithsonian Inst, Washington* (Dibner Colln)
Letters to J Collins, H Oldenburg etc, drafts of the *Tractatus de quadratura curvarum* and other notes. *Private*
Notebook 1659, papers (9) on the coinage, and astronomical observations. *Pierpont Morgan Libr, New York*
Miscellaneous correspondence and papers, incl extracts from B Valentinus, observations on the ancient solar year, letters (11ff) to H Oldenburg 1676, and warrants addressed to him as Master of the Mint 1700-13. *Brit Libr* (in Add MSS 4294, 5751, 6489, 44888)
Alchemical bibliographies, notes and recipes, notes on planets and on motion, and other papers. *Stanford Univ Libr, California* [NRA 22340]
Miscellaneous papers, incl 'Tractatus de seribus infinitis et convergentibus', mathematical figures and calculations. *Edinburgh Univ Libr* (in MS Dk 1.2)
Notebook 1662-69. *Fitzwilliam Mus, Cambridge*
'Trigonometriae fundamenta'. *Lambeth Palace Libr* (MS 592)
Prophecies concerning Christ's second coming. *Andrews Univ Libr, Berrien Springs, Michigan*
Notes (5ff) on materials and apparatus for alchemy. *Bodleian Libr, Oxford* (MS Don b 15)
Correspondence (14 items) with J Locke 1690-1703. *Ibid* (in MSS Locke c 16, 24)
Correspondence (29 items) with the Revd J Flamsteed 1681-1700. *Ibid* (CCC MS 361)
Correspondence (14 items) with the Revd J Flamsteed 1704-16. *R Greenwich Observatory, Herstmonceux* (in MSS 33, 35)

▷RS Westfall, *Never at rest: a biography of Isaac Newton*, 1930, pp875-77; *Correspondence of Isaac Newton*, ed HW Turnbull *et al*, 7 vols, 1959-77; ANL Munby, 'The Keynes Collection of the works of Sir Isaac Newton at King's College, Cambridge', *Notes and Records*, x, 1953, pp40-50; HP Macomber, *Descriptive catalogue of the Grace K Babson Collection*, 1950, and *Supplement to the catalogue of the Grace K Babson Collection*, 1955

See also Flamsteed, Gregory D, Halley, Hooke, Horsley, Oldenburg, Saunderson

[429] **NORRIS, Richard** (1830-1916)
Physiologist

Correspondence and papers, incl notes on medicine (13 vols) and photography (9 vols), notes for his physiology lectures, seance records, and MSS of his articles on spiritualism. *Birmingham Univ Libr* (MSS 15/1-212) [NRA 16930]

[430] **OLDENBURG, Henry** (?1615-1677)
FRS, man of science

Commonplace book 1654-61 and correspondence 1653-77, incl letters addressed to him, drafts (2 vols) of his letters, and 94 letters to R Boyle. *Royal Soc*
Miscellaneous correspondence and papers, incl catalogue of his books 1670 and his translation 'Of the search of truth' 1675. *Brit Libr* (Add MSS 4255-4476 *passim*)
Correspondence and papers rel to double-bottomed ships 1663-75. *Private* [NRA 23344]
Letters (50) to M Lister 1671-76. *Bodleian Libr, Oxford* (in MS Lister 34)
Letters (30) to S Hartlib 1656-60. *Sheffield Univ Libr* (50 H/39/3) [NRA 23591]
Letters to J Hevelius 1663-77. *Bibl de l'Observatoire, Paris*
Correspondence with Sir I Newton 1672-77. *Cambridge Univ Libr* (MS Add 3976)

▷*Correspondence of Henry Oldenburg*, ed AR and MB Hall, 1965-

See also Flamsteed, Lister M, Moray, Newton I, Wallis J, Willoughby

[431] **ORMEROD, George Wareing** (1810-1891)
Geologist

Geological register of notes and sections of the rocks of Lancashire and south Devon 1870-90. *Inst of Geological Sciences* [NRA 18675]
Letters (488ff) from members of his family, geologists and antiquaries 1824-84. *Bodleian Libr, Oxford* (MS Eng lett d 220)

[432] **OWEN, Sir Richard,** KCB (1804-1892) FRS, naturalist

Papers and correspondence, incl 15 pocket notebooks 1830-39, diaries of his visits to Egypt 1869 and Scotland 1872, notes (3 vols) for lectures 1828-64, papers rel to the British Museum, notes for and MSS of his *Memoir on the pearly nautilus* (1832), annotated copies of his published works, and scientific correspondence (27 vols) of Owen and W Clift. *Brit Mus (Nat Hist) General Libr*

Papers and correspondence 1826-88, incl catalogues and reports rel to the Hunterian Museum collections, palaeontological and zoological MSS, notes and drawings, drafts of his published works, and lectures on comparative anatomy etc. *R Coll of Surgeons, London*

Papers and correspondence, incl notes and lectures, mainly on anatomy and natural history 1828-80, MSS of his published works, personal correspondence with his family and others 1838-89 and letters collected by CD Sherborn. *Brit Libr* (Add MSS 33348, 34406-07, 39954-55, 42579-82, 49978)

Letters addressed to him 1838-89. *Cambridge Univ Libr* (MS 5354)

Miscellaneous letters (11) addressed to him 1834-55. *St Andrews Univ Libr*

Synopsis of a lecture course 1857, correspondence (*c*125 items) 1837-89, and miscellaneous letters to other scientists. *American Philosophical Soc, Philadelphia*

Miscellaneous correspondence and papers, incl drawings of fossil invertebrates and 34 letters to Sir RI Murchison. *Geological Soc of London*

Letters to officers of the Linnean Society, with a report on GC Holland's paper on hibernation 1856. *Linnean Soc*

Notes etc rel to the sea serpent. *Brit Mus (Nat Hist) Zoology Dept*

Annotated copy of Darwin's *Origin of species* (1859). *Shrewsbury School*

Correspondence (89ff) with Sir HW Acland 1844-90. *Bodleian Libr, Oxford* (MSS Acland d 64-200 *passim*) [NRA 22893]

Correspondence (80 items) with Sir GG Stokes 1858-89. *Cambridge Univ Libr* (in MS Add 7656)

Correspondence with William Blackwood & Sons 1851-87. *Nat Libr of Scotland* (MSS 4095-4507 *passim*)

Correspondence (44 items) with E Chadwick 1842-89. *University Coll London* [NRA 21653]

Letters (20) to the Revd W Buckland 1834-49, and 11 to Sir JFW Herschel 1844-62. *Royal Soc*

Letters (18) to Sir WJ Hooker 1839-63. *R Botanic Gardens, Kew*

Letters (18) to Sir HT De la Beche. *Nat Mus of Wales Geology Dept, Cardiff*

Correspondence (12 items) with Sir R Peel 1842-48. *Brit Libr* (Add MSS 40518-40600 *passim*)

See also Bennett, Buckland, Clift, Flower, Gould, Henslow, Hunter J

[433] **PAGET, Sir James,** 1st Bt (1814-1899) FRS, surgeon

Papers, incl 41 pocket diaries 1856-96, memoirs, reports and notes on his experiments, observations and lectures, case notes, biographical notes on students at St Bartholomew's Hospital 1843-59, drawings, and MSS of his published and unpublished works. *R Coll of Surgeons, London*

Letters (75) addressed to him 1841-94 and related correspondence and papers. *American Philosophical Soc, Philadelphia*

Index to references for his medical biographies *c*1844; testimonials 1854, 1868; letters (6) to H Pitman 1873-77. *R Coll of Physicians, London*

Letters (161ff) to Sir HW Acland and his daughter 1874-98. *Bodleian Libr, Oxford* [NRA 22893]

Letters (47) to the editors of the *Biographical dictionary* of the Society for the Diffusion of Useful Knowledge 1841-44. *University Coll London*

Letters (11) to TH Huxley 1853-95. *Imperial Coll, London*

[434] **PARSONS, Sir Charles Algernon,** OM, KCB (1854-1931) FRS, engineer

Steam turbine patent 1884; scale drawing of turbo-electric generator 1890; notebook containing calculations on parallel flow steam turbines 1897. *Inst of Mechanical Engineers* [NRA 9515]

Letters (*c*50) to the Earls of Rosse 1905-27. *St John's Coll, Cambridge*

[435] **PASCOE, Francis Polkinghorne** (1813-1893) Zoologist

Entomological scrapbook. *Hope Libr, University Mus, Oxford*

[436] **PATTINSON, Hugh Lee** (1796-1858) FRS, chemist

Correspondence and papers 1818-57, incl commonplace book, memoranda, letter books (6 vols), notes on experiments and for lectures. *Wellcome Hist Medical Libr* (MSS 3800-13)

[437] **PEACH, Benjamin Neeve** (1842-1926)
FRS, geologist

Correspondence 1862-1925 and papers 1880-1920, incl papers on fossil arthropods and on the glaciation and geology of Scotland, comments on a paper by EB Bailey, and horizontal sections through Scotland. *Inst of Geological Sciences* [NRA 18675]

See also Geikie A, Geikie J

[438] **PEACOCK, Revd George** (1791-1858)
FRS, mathematician

Correspondence (*c*530 items) *c*1820-54. *Trinity Coll, Cambridge* (MS Add b 42; Whewell Papers)
Letters (39) from J Tate 1809-43. *Richmond School, N Yorks*
Correspondence (75 items) with Sir JFW Herschel 1816-57; letters (15) to Sir JW Lubbock 1830-42. *Royal Soc*
Letters (45) to C Babbage 1816-50. *Brit Libr* (Add MSS 37182-37201 *passim*)
Letters (16) to Sir JFW Herschel 1822-58. *Texas Univ, Austin*
Correspondence (16 items) with Sir GB Airy 1844-58. *R Greenwich Observatory, Herstmonceux* (in MSS 938-47)
Letters (10) to Sir GG Stokes *c*1850-54. *Cambridge Univ Libr* (in MS Add 7656)

[439] **PEARSALL, William Harold** (1891-1964)
FRS, botanist and ecologist

Papers and correspondence (*c*20 boxes), incl field and laboratory notes and reports, papers rel to his study tours in Europe and Africa and his editorial, administrative and committee work, and notes for lectures. *Cumbria RO, Kendal*
Papers *c*1907-1964, incl field notes (75 vols) made in Europe and Africa, notes on lakes in the Lake District and on experiments, with his father, on the effects of adding nutrients to their waters 1907-22, and papers on forestry, land use, crop productivity, plant ecology etc. *Freshwater Biological Association, Ambleside*

[440] **PEARSON, Karl** (1857-1936)
FRS, mathematician and biologist

Correspondence and papers 1867-1936, incl notes of lectures attended and given 1877-1930, MSS and correspondence rel to his *Life, letters and labours of Francis Galton* (1914-30), the journal *Biometrika* 1900-36 and other published works, working papers rel to mathematics and eugenics, especially tuberculosis, albinism, insanity and his wartime research 1914-18, 3 commonplace books 1877-79, and notes on German literature. *University Coll London* [NRA 18115]
Correspondence with Sir F Galton 1893-1911. *Ibid* [NRA 19968]
Correspondence with the Eugenics Society 1921-32. *Contemporary Medical Archives Centre, Wellcome Inst for the Hist of Medicine* [NRA 24905]
Letters (21) to O Browning 1877-84. *Hastings Public Libr*
Miscellaneous letters (80). *Royal Soc*

See also Bateson, Forsyth, Lankester, Larmor, Turner HH

[441] **PELL, Revd John** (1611-1685)
FRS, mathematician

Correspondence and papers *c*1624-83, incl treatises, notes and collections rel to mathematics, logic, linguistics, shorthand, music, astronomy, chronology, theology etc (35 vols), sermons and prayers (1 vol), diplomatic papers, accounts and family papers. *Brit Libr* (Add MSS 4278-4476 *passim*)
Diplomatic correspondence and papers mainly rel to Switzerland 1648-58. *Ibid* (Lansdowne MSS 745-55)
Cypher book. *Worcester Coll, Oxford* (MSS 5.4) [NRA 10396]
Papers (3) submitted to the Royal Society. *Royal Soc*

See also Collins, Hartlib

[442] **PENGELLY, William** (1812-1894)
FRS, geologist

Notes (1 vol) on his explorations of Kent's Cavern, Torquay 1872-80. *Private*
Report of the Kent's Cavern Committee. *Royal Soc*
Catalogue of his fossil collection. *Brit Mus (Nat Hist) Palaeontology Dept*
Correspondence (19 items) with Sir GG Stokes 1861-63. *Cambridge Univ Libr* (in MS Add 7656)

[443] **PENNANT, Thomas** (1726-1798)
FRS, naturalist

Correspondence and papers, incl journals, notes on natural history, topography, genealogy etc, drawings, drafts and annotated copies of his published works, and personal, family and business correspondence. *Nat Libr of Wales* (MSS 2521-98, 4488, 4878, 5500-02, 9674, 12706-20, 21692)
Correspondence and papers, incl journals, notes on natural history, topography, antiquities, genealogy etc, drawings, materia for and annotated proofs of his published works, publication account book, local and official papers, and personal and family

correspondence. *Warwick County RO* [NRA 23685]

Journals 1765-66, 1786-88, and some miscellaneous correspondence and papers 1765-93. *Clwyd RO, Hawarden* (D/NA/839-55) [NRA 24326]

Description (22 vols) of an imaginary world tour *c*1793-98, published in part as *Outlines of the globe* (1798-1800). *Nat Maritime Mus, Greenwich*

Papers (5) submitted to the Royal Society. *Royal Soc*

His working copies of *British zoology*, 4th edn (1776-77) and *Arctic zoology*, 2nd edn (1792). *Brit Mus (Nat Hist) Zoology Dept*

Catalogues of his collections of fossils (1 vol) and minerals (2 vols). *Ibid, Mineralogy Dept*

Catalogue of his collection of minerals (1 vol). *Ibid, Palaeontology Dept*

Catalogue of 49 minerals and fossils given by him to the Ashmolean Museum 1759. *Ashmolean Mus, Oxford* (in AMS 4)

Letters (29) from LT Gronovius 1761-68 and EAW von Zimmermann 1778-96. *Brit Libr* (Add MS 52274)

Letters (283) to G Paton 1771-97. *Nat Libr of Scotland* (Adv MS 29.5.5) [NRA 23720]

Letters (*c*25) to Hester Lynch Piozzi and T Lloyd 1793-98, with MS about Vasco da Gama and Albuquerque. *John Rylands Univ Libr, Manchester* (Eng MSS 575, 892)

Letters (24) to C Linnaeus 1755-74; letters (22) to Sir JE Smith 1791-98. *Linnean Soc*

Correspondence (20 items) with Sir J Banks 1767-98. *R Botanic Gardens, Kew*

Letters to Sir J Banks 1773-93. *Mitchell Libr, Sydney* (MS 743)

Letters to the Revd G Ashby. *Suffolk RO, Bury St Edmunds* (in E 2/22/1-2)

See also Da Costa, Low, White

[444] **PERKIN, Sir William Henry** (1838-1907)
FRS, chemist

Papers and correspondence, incl laboratory notebook 1890-93, paper on cyanacetate of ethyl, notes on chemistry lectures attended 1851-52, and autobiographical papers. *Private* [NRA 19367]

School notebooks on arithmetic 1850 and chemistry 1851; notes (1 vol) on chemistry lectures attended 1855 and on his work on the aniline dye mauve 1856. *City of London School* [NRA 19367]

[445] **PETAVEL, Sir Joseph Ernest,** KBE (1873-1936)
FRS, physicist

Reports and other papers (2 files) rel to the St Louis International Exhibition 1904, notes (1 vol) for lectures on chemistry and physics given by him at St Louis, and slides of his drawings of graphs and apparatus. *Royal Inst*

Correspondence with GE Hale 1906-08. *California Inst of Technology, Pasadena*

[446] **PETIVER, James** (1663-1718)
FRS, botanist and entomologist

Correspondence and papers, incl medical journal (7 vols) 1687-1710, diary 1688-93, medical prescriptions, plant catalogues, drawings of birds, fishes and shells, annotated copies of botanical publications, and other papers rel to medicine and natural history. *Brit Libr* (Sloane MSS 1475-4070 *passim;* Add MSS 5267, 5291; 433.g.16; 968.f.8; 969.f.19-20)

Plant lists; annotated copies of J Ray's *Observations, topographical, moral and physiological* (1673) and of drawings of Cape of Good Hope plants. *Brit Mus (Nat Hist) Botany Dept*

Paper submitted to the Royal Society. *Royal Soc*

See also Blair, Bobart, Bradley R, Buddle, Dale S, Ray, Richardson R, Sherard

[447] **PETTY, Sir William** (1623-1687)
FRS, political economist

Correspondence and papers, incl letter book 1672-75, personal, official and legal papers, medical lectures given by him, Irish population returns (30 vols) *c*1659, maps and papers rel to his 'Down Survey', literary MSS, treatises on navigation and shipbuilding, and other published and unpublished writings on political, economical and military matters incl applied science and inventions. *Private* [NRA 23344]

Letter books 1666-83 and account book 1654-57. *Osler Libr, McGill Univ, Montreal* (Osler MS 7612)

MS of his 'Down Survey'. *Bibl Nat, Paris*

Papers and drawings rel to the double-bottomed ship. *Bodleian Libr, Oxford* (MS Lyell empt 32)

Papers (5) submitted to the Royal Society; letters (23) 1662-77. *Royal Soc*

[448] **PHILLIPS, John** (1800-1874)
FRS, geologist

Correspondence (*c*2,500 items) 1825-74 and papers, incl field notebooks, meteorological and geological notes, drawings of fossils, lecture scripts, catalogue of his library, and museum fossil collection day books. *Geology Dept, Oxford*

Autobiographical and literary papers. *London Univ Libr* (MS 517)

Travel journals 1829-30. *Magdalen Coll, Oxford*

Description of a tour in Scotland, incl geological observations, 1826. *Mitchell Libr, Glasgow*

Catalogues of fossil and mineral collections from 1823. *Yorkshire Mus, Geology Dept, York*

Report on the probable incidence of coal and other minerals around Lancaster 1837. *Sedgwick Mus of Geology, Cambridge*

Letters (35) to JD Forbes 1831-66. *St Andrews Univ Libr* [NRA 13132]

Letters (31) to Sir RI Murchison. *Geological Soc of London*

Letters (30) to Sir HT De la Beche. *Nat Mus of Wales Geology Dept, Cardiff*

Letters (12) to Sir GG Stokes 1847-73. *Cambridge Univ Libr* (in MS Add 7656)

Letters (12) to T Allis 1832-64. *York City Archives*

[449] **PHILLIPS, William** (1822-1905)
Botanist

Correspondence with mycologists, notebooks, a monograph on cyphellae, determinations and drawings of fungi, and other papers. *Brit Mus (Nat Hist) Botany Dept*

[450] **PILGRIM, Henry Guy Ellcock** (1875-1943)
FRS, palaeontologist

Notes (9 vols) on fossil mammals. *Brit Mus (Nat Hist) Palaeontology Dept*

[451] **PITCAIRN, Archibald** (1652-1713)
Physician

Correspondence (10 items) with R Gray 1694-1710, catalogue of books, comments on *Doctor Cockburn's solution of his problem* (1705), narrative of R Lindsay's appearances to Pitcairn in dreams, and other papers. *Brit Libr* (in Sloane MSS 3198, 3216)

[452] **PLAYFAIR, Lyon,**
1st Baron Playfair of St Andrews (1818-1898)
FRS, chemist

Correspondence (*c*1,150 items), diplomas, appointments and awards. *Imperial Coll, London*

Paper with JP Joule on atomic volumes and specific gravity; 60 letters addressed to him 1842-54. *Manchester Literary and Philosophical Soc*

Notes for a lecture on the distillation of coal 1862; correspondence with J Young 1845-82. *Andersonian Libr, Strathclyde Univ, Glasgow* [NRA 9394]

Correspondence (66 items) with Sir E Chadwick 1842-89. *University Coll London* [NRA 21653]

Letters (66) to Sir AC Ramsay 1841-43. *Nat Libr of Wales* (MS 11574)

Letters (41) to J von Liebig 1840-73. *Bavarian State Libr, Munich* (Liebigiana 57-58)

Correspondence (95ff) with WE Gladstone 1859-96. *Brit Libr* (in Add MSS 44280, 44787)

Correspondence (20 items) with Sir R Peel 1842-50. *Ibid* (Add MSS 40518-40603 *passim*)

Correspondence (17 items) with C Babbage 1851-65. *Ibid* (Add MSS 37194-99 *passim*)

Letters (13) to JD and G Forbes 1859-73. *St Andrews Univ Libr* [NRA 13132]

Correspondence (10 items) with Earl Granville 1853-57. *Public Record Office* (in PRO 30/29) [NRA 8654]

Letters (10) to TH Huxley 1861-97. *Imperial Coll, London*

Letters (10) to Sir HT De la Beche. *Nat Mus of Wales Geology Dept, Cardiff*

Miscellaneous letters (19). *Royal Inst*

See also Brewster, Buckland, Dewar, Graham, Huxley TH, Joule, Mercer

[453] **PLUKENETT, Leonard** (1642-1706)
Botanist

Correspondence 1672-1700 and papers, incl plant catalogues and 'De rebus astrologicis' 1661. *Brit Libr* (Sloane MSS 1508-4067 *passim*)

Botanical drawings and annotated copies of his published works. *Brit Mus (Nat Hist) Botany Dept*

Annotated copy of J Ray's *Catalogus plantarum Angliae* (1670). *Brit Libr* (968.f.2)

[454] **POGSON, Norman Robert** (1829-1891)
Astronomer

Variable star observations (7 vols) and charts (100); papers submitted to the Royal Astronomical Society 1854, and 9 letters to its officers 1851-84; letters to JR Hind 1850-54. *R Astronomical Soc*

Notebooks (7) containing reductions of his ring micrometer observations 1852-58. *History of Science Mus, Oxford* (MSS Radcliffe 57-63) [NRA 9532]

[455] **POLE, William** (1814-1900)
FRS, engineer

Journal of his voyage to Bombay 1844. *Brit Libr* (Add MS 42557)

Cantata and mass composed for his B Mus and D Mus degrees 1860, 1867. *St John's Coll, Oxford* (MSS 232-33)

Testimonial and other letters (100) addressed to and concerning him 1835-75. *Hove Central Libr* [NRA 10102]

Correspondence (18 items) with Sir GG Stokes 1856-97. *Cambridge Univ Libr* (in MS Add 7656)
Letters (12) to Sir JFW Herschel 1856-69. *Royal Soc*

[456] **POND, John** (1767-1836)
FRS, astronomer

Papers and correspondence 1811-36, incl meteorological journals, observations and reductions, computations, star ledgers and index errors. *R Greenwich Observatory, Herstmonceux* (MSS 304-528, 1234-37) [NRA 22822]
Paper submitted to the Royal Astronomical Society 1832; letters (8) to Sir W Herschel 1795-1811. *R Astronomical Soc*

[457] **POTT, Percivall** (1714-1788)
FRS, surgeon

MS of his *Treatise on the hydrocele* (2nd edn, 1767) 1766; student notes (3 sets) of his surgical lectures 1767-77. *R Coll of Surgeons, London* [NRA 9521]
Lectures on surgery 1775-88. *R Coll of Physicians, London*

[458] **POULTON, Sir Edward Bagnall** (1856-1943)
FRS, zoologist

Entomological diaries, notes and correspondence. *Hope Libr, University Mus, Oxford*
Notes and correspondence rel to his presidential address to the Linnean Society 1913-14. *Linnean Soc*
Letters (11) to F Galton 1888-1908. *University Coll London*
Correspondence with HH Druce. *R Entomological Soc*

See also Bateson

[459] **POWELL, Revd Baden** (1796-1860)
FRS, mathematician and physicist

Scientific correspondence *c*1836-1858. *History of Science Mus, Oxford* (MSS Mus 37, Gunther 9-10) [NRA 9532]
Letters (48) to Sir JW Lubbock 1836-55; miscellaneous correspondence (*c*60 items). *Royal Soc*
Correspondence (67 items) with Sir GG Stokes 1846-56. *Cambridge Univ Libr* (in MS Add 7656)
Letters (13) to JD Forbes 1832-42. *St Andrews Univ Libr* [NRA 13132]
Letters (12) to officers of the Society for the Diffusion of Useful Knowledge 1832-41. *University Coll London*

[460] **POWELL, Cecil Frank** (1903-1969)
FRS, physicist

Papers 1916-69, incl autobiographical writings, papers rel to his published works and speeches, and 105 notebooks 1920-69 containing records of his experimental work and of lectures and conferences attended, with drafts of his articles, lectures, speeches and letters. *Bristol Univ Libr* [NRA 17590]

[461] **POWER, Henry** (1623-1668)
FRS, physician and naturalist

Correspondence 1646-68 and papers, incl memoranda (7 vols) 1657-68, catalogue of his books 1664, and medical, chemical, astronomical, botanical and other notes and essays. *Brit Libr* (Sloane MSS 496-4067 *passim*)
His annotated copy of his *Experimental philosophy* (1664). *Ibid* (537.h.1)
Paper submitted to the Royal Society. *Royal Soc*

[462] **POWYS, Thomas Littleton,**
4th Baron Lilford (1833-1896)
Ornithologist

Correspondence (3 vols) with ACLG Günther 1868-96. *Brit Mus (Nat Hist) General Libr*
Fragment of unpublished work on the birds of Spain *c*1891. *Blacker-Wood Libr, McGill Univ, Montreal*

[463] **PRESTWICH, Sir Joseph** (1812-1896)
FRS, geologist

Field notebooks (48) rel to the British Isles 1832-90, memoirs and other papers. *Geological Soc of London*
Catalogue of specimens collected from the drift and tertiary beds 1834-76, bibliography of tertiary and quaternary geology, Ordnance Survey maps geologically coloured by him, well sections, underground temperatures and miscellaneous letters from him (8) 1877. *Inst of Geological Sciences* [NRA 18675]
Museum fossil collection day books. *Geology Dept, Oxford*
Letters (24) from H Falconer 1858-63. *Moray District RO, Forres* [NRA 20690]
Letters (21) to Sir GG Stokes 1860-85. *Cambridge Univ Libr* (in MS Add 7656)

See also Geikie A

[464] **PRIESTLEY, Revd Joseph** (1733-1804)
FRS, natural philosopher

Papers and correspondence, incl an essay on air, and *c*250 letters to the Revd T Lindsey 1766-1803. *Dr Williams's Libr, London* [NRA 13168]

Sermons (49) in shorthand 1760-83; notebook explaining Annet's shorthand system 1790; hymn *c*1790. *Manchester Coll, Oxford* [NRA 19870]
Drafts (2) of his autobiography; other papers and correspondence (125 items). *University of Pennsylvania Libr, Philadelphia*
Papers and correspondence, incl a contemporary copy of his notebook on experiments rel to air and water 1783, correspondence (25 items) with B Franklin 1766-82 and 41 letters to J Vaughan 1791-1800. *American Philosophical Soc, Philadelphia*
Papers (19) submitted to the Royal Society 1768-91, correspondence (16 items) with J Canton 1766-71 and letters (31) to J Wedgwood 1780-92. *Royal Soc*
Miscellaneous papers and correspondence from 1762. *Dickinson Coll, Carlisle, Pennsylvania*
Biographical chart of the world from 1200 BC to 1750 AD, 1765. *Bedfordshire RO* (DDX 21/581)
Correspondence (66 items) with J Wilkinson 1790-1802. *Warrington Public Libr*
Letters (20) to R Price. *Bodleian Libr, Oxford* (in MS Eng misc c 132)
Letters (11) mainly to BS Barton 1795-1803. *Hist Soc of Pennsylvania, Philadelphia*

▷RE Schofield, *A scientific autobiography of Joseph Priestley (1733-1804): selected scientific correspondence edited with commentary*, 1966

See also Withering

[465] **PRINGLE, Sir John,** Bt (1707-1782)
FRS, physician

Medical observations (10 vols) *c*1752-1777; formulae of prescriptions (2 vols); notes on various diseases and general rules for health (1 vol); commentaries on selected aphorisms of Hippocrates; pharmaceutical correspondence 1771-72. *R Coll of Physicians, Edinburgh* [NRA 16015]
Papers (10) submitted to the Royal Society. *Royal Soc*
Interleaved and annotated copy of T Sydenham's *Processus integri* 1750. *R Coll of Physicians, London*
Letters (83) to A von Haller 1760-77. *Berne Public Libr*
Miscellaneous letters (50) 1766-80. *Göttingen Univ Libr*
Letters (25) to Sir J Hall 1748-60. *Private* [NRA 22745]

[466] **PROUT, William** (1785-1850)
FRS, physician and chemist

Papers 1809-*c*1840, incl unpublished essays on the history of physic 1809 and the faculty of hearing 1810, notes for his lectures 1813-15, and papers rel to his Bridgewater treatise *c*1840. *Wellcome Hist Medical Libr* (MSS 4011-19)
Experimental notebooks (2); miscellaneous correspondence. *Private* [NRA 9764]
Paper describing an acid principle published in *Philosophical Transactions*; miscellaneous correspondence (14 items) 1827-36. *Royal Soc*

[467] **PULTENEY, Richard** (1730-1801)
FRS, botanist

Correspondence and papers, incl letters (5 boxes) to him and his wife 1747-1801, commonplace book *c*1746-1749, plant catalogues, MS and corrected proofs of his *General view of the writings of Linnaeus* (1781), annotated copy of W Hudson's *Flora Anglica* (1762), and 35 letters to Sir JE Smith 1786-1801. *Linnean Soc*
'Flora Anglica abbreviata'; catalogue of English plants and their first discoverers; catalogue of plants around Loughborough. *Brit Mus (Nat Hist) Botany Dept* (Banksian MS 90)
Letters (1 vol) addressed to him 1776-1800. *Ibid, General Libr*
Papers rel to testacea. *Ibid, Zoology Dept*
Papers (5) submitted to the Royal Society. *Royal Soc*
Catalogue of plants around Loughborough 1747. *Leicestershire RO*
Woodcuts from L Fuchs, *De stirpium historia commentarii* (1545), arranged and annotated by him. *History of Science Mus, Oxford* (MS Gunther 53)

See also Hope J, Lambert, Lightfoot

[468] **PUNNETT, Reginald Crundall** (1875-1967)
FRS, geneticist

Record books of experimental work, with W Bateson, on lathyrus from 1903, poultry and rabbits 1931-37. *Genetics Dept, Cambridge*
Papers on recessive black in poultry and on conception, notes, poems, and miscellaneous correspondence 1927-64. *Royal Soc*

See also Bateson

[469] **QUAIN, Richard** (1800-1887)
FRS, surgeon

Letters (65) 1831-56; student notes of his lectures 1837-38. *University Coll London* (College Corresp; MS Add 238)
Miscellaneous letters *c*1874-87. *R Coll of Physicians, London*

[470] **RAMSAY, Sir Andrew Crombie** (1814-1891)
FRS, geologist

Correspondence and papers, incl field notebooks and diaries (*c*40 vols) 1841-77, lecture notes and drawings, and scientific and personal correspondence. *Imperial Coll, London*

Correspondence and papers, incl autobiographical notes, letters addressed to him as director general of the Geological Survey 1871-81, 396 letters to WT Aveline 1845-83 and 75 to T Reeks 1854-79. *Inst of Geological Sciences* [NRA 18675]

Geological and family correspondence 1841-91. *Nat Libr of Wales* (MSS 7780-93, 9634-42, 11574-93)

Letters (100) to Sir HT De la Beche. *Nat Mus of Wales Geology Dept, Cardiff*

Correspondence (33 items) with Sir C Lyell 1846-74; miscellaneous correspondence (20 items) 1843-79. *Edinburgh Univ Libr* (MS Lyell 1, MS Gen 524/1)

See also Forbes E, Geikie J, Liversidge, Playfair, Sollas, Travers

[471] **RAMSAY, Sir William,** KCB (1852-1916)
FRS, chemist

Correspondence and papers, incl 10 laboratory notebooks *c*1879-1915, notes of lectures attended and given 1870-1906, correspondence and notes (16 vols) 1868-1916, MSS of his *Elements and electrons* (1912), and lists of his own published papers 1872-1912. *University Coll London* [NRA 14260]

Correspondence (100 items) with Lord Rayleigh about argon 1887-1907. *Private*

Letters (21) to Sir OJ Lodge 1885-1912. *University Coll London* [NRA 20647]

Letters (16) to A Smithells 1890-1914. *Brotherton Libr, Leeds Univ* [NRA 20283]

Letters (15) addressed to him in connection with the *Journal of Physical Chemistry* 1906. *Cornell Univ Libraries, Ithaca, New York*

Correspondence (14 items) with Lord Rutherford 1905-09. *Cambridge Univ Libr* (in MS Add 7653)

Letters (14) to Sir GG Stokes 1880-99. *Ibid* (in MS Add 7656)

Letters (11) to Lord Kelvin 1885-93. *Ibid* (in MS Add 7342)

See also Strutt JW, Travers

[472] **RASTRICK, John Urpeth** (1780-1856)
FRS, civil engineer

Diaries, notebooks and account books, plans, correspondence and papers (818 items) 1805-55, incl 132 in-letters of Stourbridge Ironworks 1830. *London Univ Libr* [NRA 12776, 15774]

Drawings (1 vol) 1840-46. *Public Record Office* (British Transport Historical Records)

Letters (15) to the Dowlais Iron Co from Rastrick and his associates 1820-39. *Glamorgan Archive Service, Cardiff* [NRA 7863]

[473] **RAY, Revd John** (1627-1705)
FRS, naturalist

Papers and correspondence, incl 'Opera quaedam de historia naturali', index, epitome and lists from his *Synopsis methodica stirpium Britannicarum* (1690), observations on the comet of 1664, description of scoter 1685, 105 letters to Sir H Sloane 1681-1705 and 19 to J Petiver 1701-02. *Brit Libr* (Sloane MSS 3322-4100 *passim;* in Add MS 4811)

Miscellaneous papers incl notes on the horseleech and other insects, annotated proofs of plates for the *Historia piscium* (1686-85) begun by F Willoughby and completed by Ray, and extracts rel to Willoughby family history. *Nottingham Univ Libr* (in Mi/LM 13, 23-25, LP 36)

Annotated copy of his *Catalogus plantarum Angliae*, 2nd edn (1677). *Brit Libr* (968.f.7)

Correspondence (114 items) with M Lister 1663-89, Sir T Robinson 1683-94 and others. *Brit Mus (Nat Hist) Botany Dept*

Letters (40) to E Lhuyd 1689-1703. *Bodleian Libr, Oxford* (in MSS Ashmole 1817a, Eng hist c 11)

Letters (21) to P Courthope 1658-74. *East Sussex RO, Lewes* (Danny Archives)

Letters (14) to J Aubrey 1676-95, with notes in Aubrey's rough draft of 'The natural historie of Wiltshire, 1685'. *Bodleian Libr, Oxford* (in MSS Aubrey 1-2, 13)

Letters (14) to the Royal Society 1670-97. *Royal Soc*

▷MA Welch, 'Francis Willoughby, FRS (1635-1672)', *J Soc Bibliog Nat Hist*, vi, pt 2, 1972, pp71-85

See also Baxter, Dale S, Douglas, Lhuyd, Lightfoot, Petiver, Plukenett, Sherard, Willoughby

RAYLEIGH, Barons, see Strutt

[474] **READE, Thomas Mellard** (1832-1909)
Geologist

Correspondence, mainly geological, containing 1,938 letters addressed to him 1870-1909, copies of his own letters (4 vols and 27 items) and 41 original letters to P Holland. *Liverpool Univ Libr* (MS 99) [NRA 9512]

[475] **RICARDO, Sir Harry Ralph** (1885-1974)
FRS, engineer

Papers, incl lectures, 1918-55. *Churchill Coll, Cambridge*

[476] **RICHARDSON, Sir Benjamin Ward** (1828-1896)
FRS, physician

Correspondence and papers *c*1852-86, incl case book, lectures on physiology, essays on diseases of the foetus, on prolonging life, and on J Black, and draft of his unfinished autobiography *Vita medica* (1897). *R Coll of Physicians, London*

Notes for his lectures on scrofula and diseases of the eye and chest *c*1855. *Wellcome Hist Medical Libr* (MS 4204)

Letters (59) to Sir E Chadwick 1876-90. *University Coll London* [NRA 21653]

[477] **RICHARDSON, Sir Owen Willans** (1879-1959)
FRS, physicist

Correspondence and papers (32,300 items), incl experimental notebooks from 1899, patents, drafts of his published works, and papers rel to cathode rays, chemical warfare research 1920, the Radio Research Board 1920-24, and the National Physical Laboratory 1946-47. *Texas Univ, Austin* [NRA 22337]

Correspondence (29 items) with N Bohr 1914-22. *Niels Bohr Inst, Copenhagen*

See also Appleton, Barkla, Bragg WH, Jeans, Rutherford, Thomson JJ

[478] **RICHARDSON, Richard** (1663-1741)
Botanist

'Deliciae hortenses' (2 vols) 1696; 'Directions for Welsh plants'; correspondence 1690-1741. *Bodleian Libr, Oxford* (MSS Radcliffe Trust c 1-11, d 1)

Botanical notes and correspondence, incl 55 letters to Sir H Sloane 1702-40 and 13 to J Petiver 1702-13. *Brit Libr* (Sloane MSS 1968-4065 *passim;* in Add MSS 4432, 4458)

Botanical notes on English and Welsh plants *c*1726. *Taylor Inst, Oxford*

Letters (34) to E Lhuyd 1690-1708. *Bodleian Libr, Oxford* (in MS Ashmole 1817a)

See also Davies, Dillenius, Lhuyd, Sloane

[479] **RIGAUD, Stephen Peter** (1774-1839)
FRS, astronomer

Correspondence and papers 1794-1835, incl mathematical tracts transcribed or composed by him (7 vols) 1794-1828, commonplace books (13 vols) 1800-39, scientific and literary notes (21 vols) 1806-39, biographical notes (8 vols) 1829-39, library catalogues, and letters addressed to him (3 vols) 1804-39. *Bodleian Libr, Oxford* (in MSS Rigaud 5-68)

Correspondence and papers, incl notes (4 vols) on trials of T Jones's meridian circle 1837-38, related correspondence and papers 1831-38, and computed occultations sent to him by the Royal Astronomical Society 1829-30. *History of Science Mus, Oxford* (MSS Radcliffe 41-44) [NRA 9532]

Papers rel to the Savilian Foundation. *Bodleian Libr, Oxford* (MSS Savile e 2-9)

Correspondence (15 items) with Sir JFW Herschel 1824-38. *Royal Soc*

See also Baily, Brewster, Maseres, Robertson A

[480] **ROBERTSON, Revd Abraham** (1751-1826)
FRS, astronomer and mathematician

Observations 1811-26. *R Greenwich Observatory, Herstmonceux*

Observations (2 vols) made at the Radcliffe Observatory 1813-15, and notes on the transit telescope. *History of Science Mus, Oxford* (MSS Radcliffe 46-48) [NRA 9532]

Notes for his lectures on astronomy *c*1813, and mathematical problems (38 items) 1794-97 transcribed by SP Rigaud. *Bodleian Libr, Oxford* (MSS Rigaud 16-19, 55)

Papers (2) submitted to the Royal Society. *Royal Soc*

Correspondence (10 items) with Sir W Herschel 1812-13. *R Astronomical Soc*

[481] **ROBERTSON, Sir Robert,** KBE (1869-1949)
FRS, chemist

Notebooks, correspondence and papers. *Public Record Office* (DSIR 26)

Biographical material (4 bundles); papers (6 boxes) rel to his attendance at international chemical congresses. *Private*

Papers and photographs rel to his work for the Royal Gunpowder Factory. *PERME Libr, Ministry of Defence, Waltham Abbey*

Correspondence and papers, mainly rel to offprints of his articles or to testimonials given by him. *Dundee Univ Libr* [NRA 19917]

Correspondence and papers (*c*35 items), mainly administrative, *c*1936-49. *Royal Inst*

Correspondence with Viscount Cherwell 1919-45. *Nuffield Coll, Oxford* [NRA 16447]

[482] **ROBINS, Benjamin** (1707-1751)
FRS, mathematician and engineer

Papers, incl commonplace book, notes on gunnery (1 vol) and other papers rel to gunnery and military affairs. *Royal Soc*

[483] **ROBINSON, Sir Robert,** OM (1886-1975)
FRS, chemist

Correspondence and papers, incl notes, drafts and correspondence with A Lapworth 1919-23. *Private*

Correspondence with Viscount Cherwell 1940-56. *Nuffield Coll, Oxford* [NRA 16447]

[484] **ROBINSON, Revd Thomas Romney** (1792-1882)
FRS, astronomer

Correspondence and papers *c*1829-1881, incl anemometer records, papers rel to nutation, magnetism, the annual motions of the earth's crust, and the visibility of the moon in total eclipses, and correspondence (*c*373 items) with Sir GG Stokes and others. *Cambridge Univ Libr* (in MS Add 7656)

Notes on astronomy (2 vols) and electricity (6 vols). *Scientific Periodicals Libr, Cambridge*

Miscellaneous correspondence and papers 1830-81, incl observations of fixed stars taken at Armagh and sent to F Baily 1830-31. *R Astronomical Soc*

Correspondence with Sir WR Hamilton 1827-77. *Trinity Coll, Dublin* [NRA 20076]

Letters (16) to C Babbage *c*1822-66. *Brit Libr* (Add MSS 37182-37201 *passim*)

Correspondence (14 items) with Sir JFW Herschel 1832-69. *Royal Soc*

See also Fairbairn, Huggins

[485] **ROBISON, John** (1739-1805)
Natural philosopher

Lectures (40 vols) on mechanics, hydrostatics, astronomy, optics, electricity and magnetism etc. *Edinburgh Univ Libr* (MSS Dc 7.1-40)

Abstract (8 vols) of his chemistry lectures at Glasgow University 1766-68; scientific commonplace book (2 vols). *St Andrews Univ Libr* (MSS 41-46, 62-63, 67)

His narrative of J Watt's invention of an improved engine, used in the case of Boulton and Watt *v* Hornblower and Maberly 1796; letters (37) to Watt 1781-1805. *Private* [NRA 22549]

Correspondence with J Watt 1768-71. *Birmingham Reference Libr* [NRA 14609]

See also Hope TC

[486] **ROLLESTON, George** (1829-1881)
FRS, physician

Papers rel to examinations 1876-80, and correspondence 1877-80. *Zoology Dept Libr, Oxford*

Archaeological papers. *Ashmolean Mus, Oxford*

Letters (37ff) to Sir HW Acland 1859-79. *Bodleian Libr, Oxford* (in MSS Acland d 65, 92, 94, 176) [NRA 22893]

Letters (30) to EB Pusey 1858-72. *Pusey House Libr, St Cross Coll, Oxford* [NRA 19807]

Letters (25) to TH Huxley 1860-81. *Imperial Coll, London*

Letters (21) to Sir WC and Sir CE Trevelyan. *Newcastle Univ Libr* [NRA 12238]

[487] **ROSCOE, Sir Henry Enfield** (1833-1915)
FRS, chemist

Student notebooks (4) kept in London and Germany 1849-55, 48 drafts of his lectures 1855-89, and 686 items of correspondence. *R Soc of Chemistry*

Notes for lectures 1857-58, letter books (2) incl reports on experiments etc 1866-83, and correspondence (2 vols) 1880-1915. *John Rylands Univ Libr, Manchester*

Report on the solar eclipse of 1870, with letters (8) to the Royal Astronomical Society's Solar Eclipse Committee. *R Astronomical Soc*

Letters (52) to Sir GG Stokes 1856-97. *Cambridge Univ Libr* (in MS Add 7656)

Letters (39) to Sir A Schuster 1879-1915. *Royal Soc*

Letters (15) to TH Huxley 1871-92. *Imperial Coll, London*

Letters (13) to HE Armstrong 1871-90. *Ibid* [NRA 11420]

Letters (10) to A Smithells 1884-1906. *Brotherton Libr, Leeds Univ* [NRA 20283]

See also Abel

[488] **ROSS, Sir Ronald,** KCB (1857-1932)
FRS, bacteriologist

Medical and literary papers and correspondence, incl his Nobel Prize lecture on malaria and MS of his *Memoirs* (1923). *R Coll of Physicians and Surgeons, Glasgow* [NRA 22540]

Papers as a medical officer in India; typescript of his *Memoirs* (1923); correspondence about the malaria mosquito, incl correspondence (*c*110 items) with Sir P Manson 1894-99. *London School of Hygiene and Tropical Medicine*

Correspondence with Sir L Rogers, mainly about malaria, with related papers, 1900-29. *Contemporary Medical Archives Centre,*

Wellcome Inst for the Hist of Medicine [NRA 24914]
Correspondence 1907-20, mainly rel to the Liverpool School of Tropical Medicine. *Liverpool Univ Archives* (TM/14/DaF, RoR) [NRA 20335]
Correspondence 1914-15. *Bodleian Libr, Oxford* (in MS Harcourt dep 506) [NRA 3679]

▷BE Beaumont, 'Sir Ronald Ross: a bio-bibliography' (Queen's Univ, Belfast thesis 1974)

[489] **ROUTH, Edward John** (1831-1907)
FRS, mathematician

Letters (25) to officers of the Royal Astronomical Society 1878-99. *R Astronomical Soc*
Correspondence (12 items) with Lord Kelvin 1882-1906. *Cambridge Univ Libr* (in MS Add 7342)

See also Darwin GH

[490] **ROXBURGH, William** (1751-1815)
Botanist

Botanical descriptions of Indian plants (23 vols); annotated transcript of his *Flora Indica* (4 vols); catalogue of botanical drawings 1791; history of hemp and flax, with drawings, 1794; correspondence and papers rel to the natural products of India (1 vol) 1789-1802; paper rel to the publication of his work. *India Office Libr and Records* (MSS Eur D 49-69, 809; Eur E 64-67; Eur F 18, 23-25; Home Misc Ser 375)
Correspondence (5 vols) *c*1789-1812; transcript of his *Flora Indica* annotated by him with R Brown (3 vols); index to his botanical MSS; 22 descriptions and drawings for his *Plants of the coast of Coromandel*, and 14 others of Malayan plants. *Brit Mus (Nat Hist) Botany Dept*
MS of his *Flora Indica* (3 vols); Sir WJ Hooker's transcript of the same (10 vols) and catalogue; another transcript (2 vols); catalogue of plants etc (1 vol). *R Botanic Gardens, Kew*
Meteorological diary kept at Fort St George, read to the Royal Society 1779. *Royal Soc*
Catalogue of Indian plants collected by him in 1792, chiefly near Madras. *R Botanic Garden, Edinburgh*
Description of swietenia febrifugia 1792. *R Coll of Physicians, Edinburgh*
Description of swietenia febrifugia *c*1793. *R Coll of Physicians, London*
Description of swietenia febrifugia. *Osler Libr, McGill Univ, Montreal*
Correspondence (41 items) with Sir J Banks 1779-1804. *Brit Libr* (in Add MSS 33977-81)
Correspondence (20 items) with T Hardwicke 1797-1811. *Ibid* (in Add MS 9869)
Letters (18) to Sir JE Smith 1791-1815. *Linnean Soc*
Letters to F Buchanan. *Scottish Record Office* (in GD 161) [NRA 8142]

See also Buchanan, Hooker WJ

RUMFORD, Count von, see Thomson, Sir B

[491] **RUSSELL, Sir Edward John** (1872-1965)
FRS, agronomist

Notes and reports on British agriculture and soils (7 boxes and 1 vol), diaries of his agricultural journeys overseas (32 vols), related notes and reports (6 boxes), notes (2 boxes) for his *History of agricultural science in Great Britain, 1620-1954* (1966) and for a book on agriculture in the USSR, and garden diary (1 vol). *Wye Coll, Kent*
Correspondence and papers *c*1886-1965, incl diaries, journals and notebooks. *Reading Univ Libr* (MS Hert 11)
Correspondence with Sir FC Bawden 1936-64. *Royal Soc* [NRA 19709]

[492] **RUTHERFORD, Ernest,**
1st Baron Rutherford of Nelson (1871-1937)
FRS, physicist

Correspondence (*c*2,400 items) and papers, incl *c*70 laboratory notebooks, other working papers, notes for lectures and speeches, and MSS of published and unpublished works. *Cambridge Univ Libr* (MS Add 7653)
Laboratory notebooks (9) 1924-26, and letters to Sir J Chadwick 1935-37. *Churchill Coll, Cambridge* (Chadwick Papers)
Joint paper with Sir OW Richardson on the structure of atomic nuclei, and 41 letters to Richardson 1929-34. *Texas Univ, Austin* [NRA 22337]
Correspondence (188 items) with N Bohr 1912-37. *Niels Bohr Inst, Copenhagen*
Letters (94) to BB Boltwood 1904-24. *Sterling Memorial Libr, Yale Univ, New Haven*
Letters (43) to Sir A Schuster 1906-30; correspondence (39 items) with Sir J Larmor 1904-25. *Royal Soc*
Correspondence with Viscount Cherwell 1914-31. *Nuffield Coll, Oxford* [NRA 16447]
Correspondence (14 items) with NV Sidgwick 1933-37. *Lincoln Coll, Oxford* [NRA 18631]
Letters (13) to GE Hale 1907-24. *California Inst of Technology, Pasadena*

▷L Badash, *Rutherford correspondence catalog*, 1974

See also Barkla, Bragg WH, Chadwick, Cockcroft, Dewar, Glazebrook, Laby, Larmor, Lodge, McLennan, Moseley, Ramsay W, Schuster, Soddy, Thomson JJ, Townsend

[493] **SABINE, General Sir Edward,** KCB (1788-1883)
FRS, geophysicist

Personal and official correspondence and papers (57 files and volumes) 1825-77, incl records of his Magnetic Department rel to the establishment of magnetic observatories overseas. *Public Record Office* (BJ 3)
Correspondence (1,841 items), memoirs (6), journal of observations made on the Ross expedition 1818; correspondence (363 items) with Sir JFW Herschel 1825-70; letters (35) to Sir JW Lubbock 1830-52. *Royal Soc*
Diary kept during the Ross expedition 1818. *West Devon Area RO, Plymouth* (920 SAB)
Papers rel to pendulum stations and longitudes 1822; correspondence (*c*11 items) with Sir GB Airy *c*1847-58. *R Greenwich Observatory, Herstmonceux* (in MSS 938-48)
Correspondence rel to the Meteorological Office and Kew Observatory. *Meteorological Office Archives, Bracknell*
Correspondence (283 items) with Sir GG Stokes and others *c*1850-76. *Cambridge Univ Libr* (in MS Add 7656)
Correspondence (103 items) with Sir JFW Herschel 1822-66. *Texas Univ, Austin*
Letters (57) to Sir WR Grove; letters (*c*40) to J Tyndall. *Royal Inst*
Letters (59) to W Sharpey 1852-65. *University Coll London* (in MS Add 227)
Correspondence (34 items) with JD Forbes 1841-58. *St Andrews Univ Libr* [NRA 13132]
Letters (26) to Sir WJ Hooker 1832-49. *R Botanic Gardens, Kew*
Letters (26ff) to Sir RI Murchison 1845-62. *Brit Libr* (in Add MS 46128)
Letters (*c*26ff) to the Marquess of Northampton. *Private* [NRA 21088]
Letters (24) to the Göttingen Academy of Sciences 1824-65. *Göttingen Univ Libr*
Letters (10) to M Faraday 1845-60. *Inst of Electrical Engineers* (Blaikley Colln)

See also Baily, Fox RW, Lloyd H, Stewart

[494] **SALAMAN, Redcliffe Nathan** (1874-1955)
FRS, physiologist

Notes on the history of the potato (11 vols), and other papers rel to his research on potato genetics, Jewish heredity, Zionism, and his family. *Cambridge Univ Libr*
Accounts for Smith End Farm, Barley, 1925-48. *Hertfordshire RO*

[495] **SALISBURY, Richard Anthony** (1761-1829)
Botanist

Botanical papers and drawings (9 vols); notes and drawings of ericaceous plants. *Brit Mus (Nat Hist) Botany Dept*
Correspondence *c*1797-1829, autobiographical notes, and MSS of his published and unpublished works. *Linnean Soc*

[496] **SALMON, Revd George** (1819-1904)
FRS, mathematician

Family papers (9 vols). *Trinity Coll, Dublin* (MSS 4738-46) [NRA 19217]
Letters (1 vol) to the Revd JH Bernard 1891-1903, and miscellaneous papers and correspondence of and rel to him collected by Bernard. *Ibid* (MSS 2384-2385a) [NRA 20074]
Letters (48) to WEH Lecky 1895-1903. *Ibid* [NRA 20228]
Letters (18) to E Dowden 1879-1900. *Ibid* [NRA 20227]
Letters (18) to TA Hirst 1858-73. *University Coll London* (London Mathematical Society Papers)

[497] **SALVIN, Osbert** (1835-1898)
FRS, naturalist

Letter book, letters to A Newton and miscellaneous papers. *Balfour and Newton Libr, Zoology Dept, Cambridge*
Field notebooks (3) rel to central America 1857-59, and list of birds 1862; correspondence, notebooks and other papers, with FD Godman, rel to their *Biologia Centrali-Americana* (1879-1915). *Brit Mus (Nat Hist) Zoology Dept*
Drawings, with FD Godman, for the insecta volumes of their *Biologia Centrali-Americana. Ibid, Entomology Dept*
Notes on the physical geography of Guatemala 1862. *R Geographical Soc*

[498] **SAMPSON, Ralph Allen** (1866-1939)
FRS, astronomer

Correspondence and papers 1910-37, rel to transit circles, time signals, clocks and chronometers, solar spectroscopy, photometry and photography, seismology, meteorology, rangefinding, optical systems design etc, and administrative records as astronomer royal for Scotland. *R Observatory, Edinburgh*
Correspondence (*c*300 items) 1896-1911 and papers as director of Durham University Observatory 1896-1910, incl notes, tables,

charts etc (8 boxes) rel to his work on Jupiter's satellites. *Durham Univ Libr*
Letters (33) to officers of the Royal Astronomical Society 1892-1900. *R Astronomical Soc*
Letters (30) to Sir J Larmor. *St John's Coll, Cambridge*

[499] **SAUNDERS, William Wilson** (1809-1879)
FRS, entomologist

Illustrated journal of voyage to Calcutta 1830, other journals 1833-40, and family correspondence. *Buckinghamshire RO, Aylesbury*
Entomological notebooks. *Hope Libr, University Mus, Oxford*
Drawings (466) and tracings (105) of British basidiomycetes; collection of drawings of other fungi. *Brit Mus (Nat Hist) Botany Dept*
Letters (82) to Sir WJ Hooker 1843-64. *R Botanic Gardens, Kew*

[500] **SAUNDERSON, Nicholas** (1682-1739)
FRS, mathematician

Treatise on fluxions; lectures on hydrostatics, optics, the tides etc. *Cambridge Univ Libr* (MSS Add 589, 2977, 3444, 6312)
Treatise on fluxions; commentary on Newton's *Principia*; lectures on mathematics 1736. *Stanford Univ Libr, California*
Commentary on Newton's *Principia*. *Babson Coll, Babson Park, Mass* (MS 454)
Lectures on hydrostatics, optics, tides, the rainbow, heat and cold, and sound, c1725-1736. *Wellcome Hist Medical Libr* (MSS 4373-74)
Lectures (3 vols) on hydrodynamics, mechanics, sound etc 1728. *Norfolk RO*
Lectures on mathematics 1731. *University Coll London* (MS Add 243)
Lectures on physics etc. *Bodleian Libr, Oxford* (MSS Rigaud 3-4)
Lectures on mechanics, hydrostatics, pneumatics, the tides, sounds, optics, microscopes, telescopes and astronomy. *Brit Libr* (Add MS 57488)

[501] **SAVILE, Sir Henry** (1549-1622)
Mathematician

Collection of mathematical treatises, lecture notes, papers and correspondence; printer's copy of his translations from Tacitus annotated by him 1591. *Bodleian Libr, Oxford* (MSS Savile 1-52; MS Eng hist d 240)

See also Rigaud

[502] **SCHONLAND, Sir Basil Ferdinand Jamieson** (1896-1972)
FRS, physicist

Papers (5 boxes) rel to the Atomic Energy Research Establishment, Harwell, 1954-60. *Churchill Coll, Cambridge*

[503] **SCHORLEMMER, Carl** (1834-1892)
FRS, chemist

MS of his unfinished history of chemistry. *John Rylands Univ Libr, Manchester* [NRA 9194]

[504] **SCHUSTER, Sir Arthur** (1851-1934)
FRS, physicist

Diary, accounts and photographs of Siam expedition to observe solar eclipse 1875; letters (16) to officers of the Royal Astronomical Society 1877-98 and its Joint Permanent Eclipse Committee 1903-11. *R Astronomical Soc*
Lectures; diary notes; photographs of the eclipse 1875. *Private*
Correspondence (3 vols). *Royal Soc*
Letters (33) to Sir J Larmor. *St John's Coll, Cambridge*
Letters (24) to Sir OJ Lodge 1884-1914. *University Coll London* (MS Add 89)
Letters (18) to Sir GG Stokes 1881-97. *Cambridge Univ Libr* (in MS Add 7656)
Correspondence (17 items) with Sir JJ Thomson. *Ibid* (in MS Add 7654)
Letters (14) to Lord Rayleigh 1880-1909. *Private*
Letters (13) to Lord Rutherford 1906-19. *Cambridge Univ Libr* (in MS Add 7653)
Correspondence (11 items) with Lord Kelvin 1882-1904. *Ibid* (in MS Add 7342)

See also Roscoe, Rutherford, Thomson JJ, Thomson W

[505] **SCOTT, Hugh** (1885-1960)
FRS, entomologist

Papers, incl journals 1899-1937, notebooks and 8 photograph albums. *Brit Mus (Nat Hist) General Libr*
Letters addressed to him 1901-56. *Ibid, Entomology Dept*
Correspondence with HStJ Philby on the topography and natural history of Arabia 1939-54. *Middle East Centre, St Antony's Coll, Oxford*

[506] **SEDGWICK, Revd Adam** (1785-1873)
FRS, geologist

Correspondence and papers, incl 42 field notebooks 1818-57, research and lecture notes, catalogues of specimens and related

correspondence, catalogue of his library and annotated maps. *Sedgwick Mus of Geology, Cambridge*
Geological correspondence and papers, and biographical materials collected by TM Hughes. *Cambridge Univ Libr* (MS Add 7652)
Journal of a tour to France *c*1816, MSS of papers read to the Geological Society of London, and *c*45 letters to Sir RI Murchison. *Geological Soc of London*
Papers rel to T Image's collection of fossils 1856-58. *Cambridge Univ Archives*
Correspondence (2 vols); letters (70) to W Whewell 1827-59. *Trinity Coll, Cambridge*
Letters (20) to Sir JFW Herschel 1820-69. *Royal Soc*
Letters (19) to C Grey 1851-69. *Dept of Palaeography and Diplomatic, Durham Univ* [NRA 6228]
Correspondence (17 items) with Sir GB Airy 1845-55. *R Greenwich Observatory, Herstmonceux* (in MSS 939-46)
Letters (11) to Lord Brougham 1826-54. *University Coll London*
Letters (10) to C Babbage 1817-68. *Brit Libr* (Add MSS 37182-99 *passim*)

See also Bateson, Clarke WB, Murchison

[507] **SEEBOHM, Henry** (1832-1895)
Ornithologist

Ornithological journals and travel diaries 1875-81. *Cambridge Univ Libr* (MSS Add 4471-76)
Travel diaries (4) 1872-82; catalogue of birds' eggs in the Museum (10 vols). *Brit Mus (Nat Hist) Zoology Dept*
Family correspondence. *Hertfordshire RO* (D/ESe) [NRA 4447]

[508] **SEELEY, Harry Govier** (1839-1909)
FRS, geologist and palaeontologist

'An outline of the osteology of . . . reptiles' (1 vol); correspondence (1 vol) with DR Kannemeyer 1891-1905; letters to AS Woodward 1892-93. *Brit Mus (Nat Hist) Palaeontology Dept*
Letters (56ff) from scientists *c*1860-1902. *Brit Libr* (in Add MS 42585)
Letters to his wife from South Africa 1889. *Private*

[509] **SEEMANN, Berthold Carl** (1825-1871)
Botanist

Papers and correspondence, incl journal of voyage of HMS *Herald* 1847-51, correspondence with G Bentham 1847-57 and 47 letters to Sir WJ Hooker 1846-60. *R Botanic Gardens, Kew*
Papers read to the Linnean Society. *Linnean Soc*

[510] **SEWARD, Sir Albert Charles** (1863-1941)
FRS, botanist and geologist

Notes (2 vols) on fossil plants in European museums 1888. *Botany School, Cambridge* [NRA 9541]
Typescript and illustrations of 'Antarctic fossil plants of the 1910 *Terra Nova* expedition' (1914). *Brit Mus (Nat Hist) Palaeontology Dept*
Correspondence about the publication of *Darwin and modern science* (1909). *Cambridge Univ Libr*
Correspondence (30 items) with FO Bower 1912-26. *Glasgow Univ Archives* [NRA 18906]

[511] **SHARPEY, William** (1802-1880)
FRS, physiologist

Letters (214) addressed to him 1852-66, mainly as secretary of the Royal Society; student notes of his anatomy and physiology lectures 1836-67. *University Coll London* (MSS Add 227, 238; Medical Sciences) [NRA 18112, 20580]
Letter book kept by him as secretary of the Royal Society 1853-55, with diplomas and certificates 1821-74. *Arbroath Signal Tower Mus* [NRA 24032]
Correspondence, testimonials etc 1829-80. *Edinburgh Univ Libr*
Correspondence (118 items) with Sir GG Stokes and others 1855-80. *Cambridge Univ Libr* (in MS Add 7656)
Miscellaneous letters (16). *Royal Soc*

See also Brodie BC the elder, Lister J, Sabine, Stokes

[512] **SHEEPSHANKS, Revd Richard** (1794-1855)
FRS, astronomer

Correspondence and papers, incl abstracts prepared for the *Monthly Notices* of the Royal Astronomical Society 1827-29, 3 reports submitted to the Society, *c*102 letters to A De Morgan *c*1842-52, 28 to HC Schumacher 1830-50, 17 to officers of the Society 1840-55, and 576 letters addressed to him. *R Astronomical Soc*
Books (16) of original comparisons of standard bars and thermometers; correspondence 1838-55, incl correspondence (94 items) with Sir GB Airy 1845-55. *R Greenwich Observatory, Herstmonceux* (MSS 926-46 *passim*)

Correspondence (142 items) with Sir JFW Herschel 1825-52; letters (12) to Sir JW Lubbock 1830-47. *Royal Soc*
Letters (34) to W Whewell 1821-54. *Trinity Coll, Cambridge*
Letters (16) to the Society for the Diffusion of Useful Knowledge 1830-38. *University Coll London*
Letters (14) to Sir JFW Herschel 1841-53. *Texas Univ, Austin*

See also Adams, Airy, Johnson MJ, Smyth

[513] **SHERARD, William** (1659-1728)
FRS, botanist

Unpublished continuation of C Bauhin's *Pinax*, plant lists, papers, drawings and correspondence. *Bodleian Libr, Oxford* (MSS Sherard 44-201, MSS Radcliffe Trust c 1-11) [NRA 6305]
Correspondence (764 items); miscellaneous papers. *Royal Soc*
Notes and observations on J Ray's *Historia plantarum* I-II (1686-88). *Brit Mus (Nat Hist) Botany Dept* (Banksian MS 80)
Notes on inscriptions in Asia Minor 1705-16. *Brit Libr* (Add MS 10101)
Letters (49) to Sir H Sloane 1690-1724, and 17 to J Petiver 1700-15. *Ibid* (Sloane MSS 4036-67 *passim*)
Copy letters (38) to him as British consul at Smyrna from the Levant Company 1703-16. *Public Record Office* (in SP 105/115-16)

See also Bobart

[514] **SHERRINGTON, Sir Charles Scott,** OM, GBE (1857-1952)
FRS, physiologist

Reports (15) on papers submitted to the Royal Society 1893-98, 25 letters to RA Granit and *c*100 other letters from him. *Royal Soc*
Correspondence and papers rel to research topics and administration. *Medical Research Council*
Papers, mainly drafts of *Goethe on nature and on science* (2nd edn, 1949) and 'Marginalia' in *Science, medicine and history*, ed EA Underwood (1953). *Liverpool Univ Libr* (MS 146)
Miscellaneous letters and papers *c*1892-1951. *Woodward Biomedical Libr, Univ of British Columbia, Vancouver*
Letters 1907-15. *Liverpool Univ Archives*
Correspondence (over 20 items) with AV Hill. *Churchill Coll, Cambridge*
Letters (16) to A Ruffini 1896-1903. *Private*

[515] **SHUCKBURGH-EVELYN, Sir George Augustus William,** 6th Bt (1751-1804)
FRS, mathematician

Astronomical observations 1774-97. *Cambridge Univ Libr* (MSS Add 4523-34)
Letters (31) to J Lloyd 1778-1801. *Nat Libr of Wales* (MS 12418)

[516] **SIBTHORP, John** (1758-1796)
FRS, botanist

Botanical and zoological correspondence and papers, incl diaries 1786-94, notebooks, lists of plants and animals, lectures on botany, notes for and MSS of his published works. *Bodleian Libr, Oxford* (MSS Sherard 208-463 *passim*) [NRA 6305]
Notebooks and papers for his posthumous *Flora Graeca* (1806-40). *Botany Dept, Oxford* (in MSS Sherard 248, 456-57)
Correspondence (66 items) with J Hawkins 1785-95. *West Sussex RO, Chichester*
Letters (10) from him to his family 1782-86. *Lincolnshire Archives Office* [NRA 0584]

See also Daubeny, Smith JE

[517] **SIDGWICK, Nevil Vincent** (1873-1952)
FRS, chemist

Correspondence and papers 1885-1952, incl travel journals 1914-31, laboratory notes and notes on his chemistry students 1895-1949, lectures 1914-51, essays 1892-96 and correspondence 1902-51. *Lincoln Coll, Oxford* [NRA 18631]
Papers 1899-1934, mainly rel to the sub-faculty of chemistry, with some notes of lectures attended and given. *History of Science Mus, Oxford* (MSS Mus 122, 131-33, 139) [NRA 18631]

See also Rutherford

[518] **SIMON, Sir Francis Eugene** (1893-1956)
FRS, physicist

Correspondence and papers 1919-56, incl laboratory notebooks and working papers 1919-34, account book 1931-33, notes for lectures and addresses *c*1930-1956, and drafts of his published works. *Royal Soc* [NRA 13900]
Correspondence with Viscount Cherwell 1933-56, and papers incl a memorandum 1940 and report 1942 on liquid hydrogen. *Nuffield Coll, Oxford* [NRA 16447]
Correspondence with Lord Blackett 1945-48. *Royal Soc* [NRA 22627]

[519] **SIMON, Sir John,** KCB (1816-1904)
FRS, sanitary reformer

Correspondence and papers, incl annotated copy of his autobiography, notes and drawings from 1838, annotated articles and essays. *R Coll of Surgeons, London* [NRA 9521]
Correspondence of Simon and his wife (67 items) with A and J Severn 1878-1902. *John Rylands Univ Libr, Manchester* (MS 1260)
Letters (19ff) to Sir HW Acland 1884-96. *Bodleian Libr, Oxford* (MSS Acland d 62-98 *passim*) [NRA 22893]

[520] **SIMPSON, Sir James Young,** 1st Bt (1811-1870)
Physician

Papers and correspondence (19 boxes), on medical, personal, family, business, archaeological and religious topics. *R Coll of Surgeons, Edinburgh* [NRA 20261]
Lectures 1837-62; case book 1836-43; notebooks; letter books (2 vols) 1845-57. *R Coll of Physicians, Edinburgh* [NRA 16015]
Papers (103 items) from 1838. *Duke Univ Medical Libr, Durham, N Carolina*
Dissertation 'On the disease of the placenta' 1835. *R Medical Soc, Edinburgh*

[521] **SLOANE, Sir Hans,** Bt (1660-1753)
FRS, physician

Correspondence and papers, incl medical treatises, chemical and medical receipts, a collection of papers rel to natural history, medicine etc, and minutes of the Royal Society 1696-1712. *Brit Libr* (Sloane MSS; Add MS 29740)
Catalogues (20 vols) of his natural history collections. *Brit Mus (Nat Hist) Botany, Mineralogy, Palaeontology and Zoology Depts*
Marriage settlement; papers rel to Jamaican plantations 1694-1733. *Lincolnshire Archives Office* (Ancaster Papers) [NRA 5789]
Papers (13) submitted to the Royal Society, and miscellaneous letters (13) 1692-1737. *Royal Soc*
Letters to R Richardson. *Bodleian Libr, Oxford* (in MSS Radcliffe Trust c 1-11)

See also Blair, Bobart, Bradley R, Collinson, Cowper, Dale S, King E, Lhuyd, Musgrave, Ray, Richardson R, Sherard, Wallis J, Woodward J

[522] **SMEATON, John** (1724-1792)
FRS, civil engineer

Civil and mechanical engineering designs (1,200 in 11 vols) 1741-92; MSS (23) of papers published in *Philosophical Transactions* 1750-88. *Royal Soc*
Letter book 1764; letters (4 vols) 1781-88; will 1792. *Inst of Civil Engineers* [NRA 16921]
Diary kept in the Low Countries 1754. *Trinity House*
Drawing for Calder bridge 1757; plans (6) for Middleton engine and engine house 1779-80; correspondence rel to Aire and Calder navigation. *Leeds City Archives Dept* [NRA 9510]
Contract with Glasgow city magistrates for construction of a lock and dam 1760. *Strathclyde Regional Archives, Glasgow* (Clyde Port Authority Records) [NRA 13871]
Plans for Cardington mill. *Bedfordshire RO* (Whitbread Colln)
Memorandum rel to a wall and landing steps at Greenwich 1787. *R Inst of British Architects* [NRA 13990]
Letters (14) to B Wilson 1744-64. *Brit Libr* (in Add MS 30094)

SMITH, Revd FJJ, see Jervis-Smith

[523] **SMITH, Sir Grafton Elliot** (1871-1937)
FRS, anatomist

MSS of lectures and articles 1934-35; fragment of autobiography; correspondence (440 items) 1896-1936. *Brit Libr* (Add MSS 56303-04)
Letters (52) to JT Wilson 1896-1936. *Australian Academy of Science, Canberra*
Miscellaneous correspondence, mainly rel to research topics. *Medical Research Council*
Correspondence with GG Campion 1921-36. *John Rylands Univ Libr, Manchester*

[524] **SMITH, Sir James Edward** (1759-1828)
FRS, botanist

Correspondence (over 3,000 items) 1781-1828 and papers, incl MSS of his published works, lectures, botanical notebooks from 1779, lists of plants and insects, diplomas 1786-1824, and accounts for the purchase and sale of books 1796-1815. *Linnean Soc*
Papers rel to his edition of J Sibthorp's *Flora Graeca*, incl part of the MS and proofs. *Bodleian Libr, Oxford* (in MSS Sherard 249-50) [NRA 6305]
Lecture on *Florae Scoticae supplementa* read to the Society for the Investigation of Natural History at Edinburgh. *Scottish Record Office* (in GD 253/144)
Correspondence (3 vols) from 1793, selected as autographs by his wife. *Beinecke Libr, Yale Univ, New Haven* (Osborn Colln, MS d 148/1-3) [NRA 18661]
Letters (123) to J Sowerby 1789-1802, and other letters to J and J de C Sowerby. *Brit Mus (Nat Hist) General Libr*

Correspondence (21 items) with J and J de C Sowerby 1816-27; correspondence (14 items) with J Hawkins 1801-15. *Botany Dept, Oxford* (in MS Sherard 248)

Letters (18) to AB Lambert 1788-1808. *Brit Libr* (in Add MS 28545)

▷WR Dawson, *Catalogue of the manuscripts in the Library of the Linnean Society of London. I. The Smith Papers*, 1934

See also Banks, Boott, Borrer, Brown R, Buchanan F, Davies, Forster, Hardwicke, Hooker WJ, Hope TC, Lambert, Martyn, Pennant, Pulteney, Roxburgh, Sowerby

[525] **SMITH, William** (1769-1839)
Geologist and engineer

Correspondence (*c*900 items) 1793-1839 and papers, incl diaries (3) 1789-1821, autobiographical notes, sketches of his acquaintances (1 vol), MSS of the unpublished parts of *Strata identified by organized fossils* and other writings, geological observations, notes, sections and maps. *Geology and Mineralogy Dept, Oxford*

'A delineation of the strata of England and Wales with part of Scotland' 1812-15, 6 coloured geological views and sections across England and Wales 1817-19, and catalogue of fossils delivered to the British Museum 1818. *Brit Mus (Nat Hist) Palaeontology Dept*

Table of strata near Bath written at his dictation 1799, and miscellaneous geological maps *c*1800-1831. *Geological Soc of London*

Miscellaneous papers on geology, agriculture, politics, statistics etc. *Scarborough Mus*

Observations on the vale of the Waveney 1806. *Fitzwilliam Mus, Cambridge* (SG Perceval Colln)

Geological section from London to Snowdon 1817. *West Sussex RO, Chichester*

Map of Essex geologically coloured by him 1824. *Essex RO, Chelmsford*

▷T Sheppard, 'William Smith: his maps and memoirs', *Proceedings of the Yorkshire Geological Society*, xix, pt 2, 1916, pp79-253

[526] **SMITHELLS, Arthur** (1860-1939)
FRS, chemist

Scientific notebooks, Indian journal, correspondence and papers (*c*800 items) 1875-1939. *Brotherton Libr, Leeds Univ* [NRA 20283]

Correspondence (280 items) with Sir GG Stokes 1882-1901. *Cambridge Univ Libr* (in MS Add 7656)

Correspondence (20 items) with Sir OJ Lodge 1893-1932. *University Coll London* (in MS Add 89)

See also Dixon, Ramsay W, Roscoe, Thorpe TE

[527] **SMYTH, Charles Piazzi** (1819-1900)
FRS, astronomer

Correspondence and papers 1833-94, incl observations and reductions, papers rel to spectroscopy, meteorology, photography, pyramids, and his experiments on Tenerife, and administrative records as astronomer royal for Scotland. *R Observatory, Edinburgh*

Notebooks and other papers mainly rel to photography. *Royal Soc of Edinburgh*

Drawings of Tenerife 1856; photographs of cloud forms (3 vols) 1892-94. *Royal Soc*

Letters, incl 62 to the officers of the Royal Astronomical Society 1846-92, and 12 to R Sheepshanks 1846-52. *R Astronomical Soc*

Correspondence (27 items) with JD and G Forbes 1846-80. *St Andrews Univ Libr* [NRA 13132]

Correspondence (17 items) with Sir GB Airy 1849-66. *R Greenwich Observatory, Herstmonceux* (in MSS 941-51)

Letters (15) to Sir D Gill 1872-79. *R Geographical Soc*

Letters (13) to J Lee 1849-53. *Edinburgh Univ Libr* (in MS Dc 3.99/6)

Correspondence (12 items) with Sir GG Stokes 1856-71. *Cambridge Univ Libr* (in MS Add 7656)

Letters (12) to Sir JFW Herschel 1842-57. *Royal Soc*

Letters (10) to DR Hay 1849-60. *Edinburgh Univ Libr* (MS Dc 2.58)

[528] **SODDY, Frederick** (1877-1956)
FRS, chemist

Papers and correspondence 1894-1956, incl laboratory notebooks (76), lectures, articles and addresses. *Bodleian Libr, Oxford* [NRA 18545]

Papers, incl notes for 23 lectures on radioactivity 1914-17. *History of Science Mus, Oxford* (MSS Mus 121, 126, 160)

Correspondence (96 items) with Lord Rutherford 1902-37. *Cambridge Univ Libr* (in MS Add 7653)

Miscellaneous letters (25). *Royal Inst* [NRA 9522]

[529] **SOLANDER, Daniel Carlsson** (1733-1782)
FRS, botanist

Papers, incl his slip catalogue (24 vols) of species of plants, descriptions and lists (9 vols) of plants collected during Cook's first voyage 1768-71, memoranda rel to Iceland 1772, and other plant descriptions, lists and indexes. *Brit Mus (Nat Hist) Botany Dept*

Papers, incl his slip catalogue (27 vols) of species of animals, descriptions of animals collected during Cook's first voyage 1768-71, descriptions of fishes collected in Iceland 1772. *Ibid, Zoology Dept*

Diaries as assistant keeper of natural history, British Museum, 1764-82, and letters addressed to him, mainly by J Ellis, 1760-63. *Brit Libr* (in Add MSS 29533, 45874-75)

Correspondence and papers, incl descriptions of 13 species of tellina and 43 insects, memoranda rel to his edition of J Ellis's *Natural history of zoophytes* (1786) and 28 letters to Ellis 1761-76. *Linnean Soc*

Notes on Linnaeus's *Flora Suecica* (1745). *Cambridge Univ Libr*

Tahitian vocabulary. *School of Oriental and African Studies, London* (MS 12023)

See also Ellis J

[530] **SOLLAS, William Johnson** (1849-1936)
FRS, geologist

Notebooks, lecture notes, *c*200 letters, and a collection of photographs of serial sections of fossil skulls. *University Mus, Oxford* (Geological Collns)

Illustrated notes on lectures by Sir AC Ramsay (1 vol); college diplomas. *Imperial Coll, London*

Letters (111ff) from scientists 1875-1927. *Bodleian Libr, Oxford* (MS Eng lett d 329)

Letters from TG Bonney 1885-1920. *University Coll London*

Section of boring through the atoll Funafuti, Ellice Is; correspondence (14 items) 1892-1926. *Inst of Geological Sciences* [NRA 18675]

[531] **SOMERVILLE, Mary** (1780-1872)
Scientific writer

Scientific, business, personal and family correspondence and papers 1806-72, incl notebooks rel to physics, astronomy, and mathematics, European diaries, drafts of her published works and of her autobiography. *Bodleian Libr, Oxford* (MSS Dep c 351-78) [NRA 9423]

Correspondence (58 items) with Sir JFW Herschel 1829-69; letters (10) to Sir JW Lubbock 1829-34. *Royal Soc*

Miscellaneous letters (34) 1832-70. *Somerville Coll, Oxford* [NRA 9423]

Letters (20) to C Babbage 1830-48. *Brit Libr* (Add MSS 37185-37201 *passim*)

Correspondence (13 items) with W Whewell 1828-48. *Trinity Coll, Cambridge* [NRA 8804]

See also Babbage, Lyell, Whewell

[532] **SORBY, Henry Clifton** (1826-1908)
FRS, geologist

Diaries (11 vols) 1859-1908, and photographic collection. *Sheffield Univ Libr* (MS 67) [NRA 20467]

Papers rel to the Royal Society 1857, 1874; experimental illustrations (1 folder); letters addressed to him (1,128) 1844-96. *Sheffield Central Libr* [NRA 9535]

Sketches of natural history specimens and scenery, with some correspondence and photographs. *Sheffield City Mus*

Letters (82) to Sir GG Stokes 1855-89. *Cambridge Univ Libr* (in MS Add 7656)

[533] **SOWERBY, James** (1757-1822)
Naturalist

Correspondence and papers, incl *c*2,760 letters addressed to him or to his son James, 123 letters from Sir JE Smith 1789-1802, accounts from 1782, drafts, notes and drawings for his publications, and annotated proof plates. *Brit Mus (Nat Hist) General Libr*

Correspondence and papers, incl a report and correspondence on dry rot in ships 1812-14, drawings of fungi (213) with related papers, drawings of plants at Grove Hill, Camberwell *c*1787 (2 vols) and for his own and other published works. *Ibid, Botany Dept*

Geological drawings and papers, and drawings for Sir JE Smith's *Plantarum icones hactenus ineditae* (1789-91), with 7 letters to Smith 1800-22. *Linnean Soc*

Paper on Kimmeridge and other coals 1808. *Geological Soc of London*

Letters (95) addressed to him or to his son James from *c*1789. *Royal Soc*

Letters (66ff) addressed to him 1793-1814; letters (12) from C Lyell 1806-12. *Nat Libr of Scotland* (MSS 3925, 10789)

Letters (*c*30) to the Revd J Dalton *c*1795-1822. *Brotherton Libr, Leeds Univ*

Letters (26) to H Davies. *Nat Libr of Wales* (in MS 6664)

Letters (14) to W Cunnington. *Wiltshire Archaeological and Nat Hist Soc, Devizes*

Correspondence with his sons 1797-1822. *Private*

▷*J Soc Bibliog Nat Hist*, vi, pt 6, 1974, pp380-559 and vii, pt 4, 1976, pp343-68

See also Smith JE

[534] **SPENCER, Leonard James** (1870-1959)
FRS, mineralogist

Working notebooks and official correspondence. *Brit Mus (Nat Hist) Mineralogy Dept*

Correspondence, diaries and papers. *Private*

[535] SPOTTISWOODE, William (1825-1883)
FRS, mathematician and physicist

List of apparatus and notes (2 vols). *Royal Inst*
Letters (78) to Sir GG Stokes 1868-83. *Cambridge Univ Libr* (in MS Add 7656)
Letters (18) to officers of the Royal Astronomical Society 1858-83. *R Astronomical Soc*
Letters (11) to TH Huxley 1866-83. *Imperial Coll, London*

[536] SPRUCE, Richard (1817-1893)
Botanist

Correspondence 1842-80 and papers, incl travel journals 1851-57, catalogues of plants of the Amazon (2 vols) 1849-55 and Andes (2 vols) 1855-56, notes for the introduction to his *Hepatics of the Amazon and Andes* (1884-85), other papers rel to mosses and hepaticae, correspondence (125 items) with Sir WJ Hooker 1842-65, and letters (66) to W Borrer 1842-48. *R Botanic Gardens, Kew*
Papers rel to his travels in S America, incl meteorological registers, Liverpool to Pará 1849, Rio Negro and the Andes 1851-62; notes on Tarapoto, Peru; Tupi vocabulary and notes on other American languages. *R Geographical Soc*
Miscellaneous correspondence, accounts and other papers, incl his field journal 1849-50. *Manchester Central Libr, Archives Dept*
List of botanical excursions 1841-63, with notes on the acclimatization of Europeans in tropical S America. *Linnean Soc*
Correspondence with D Hanbury 1856-75. *Pharmaceutical Soc of Great Britain*
Drawings of scenery and natives in Amazonian Brazil *c*1850. *Royal Soc*

[537] STAINTON, Henry Tibbats (1822-1892)
FRS, entomologist

MS and drawings for his *Natural history of the tineina* (1855-73), letters addressed to him 1852-92, and correspondence (4 vols) with PC Zeller 1869-83. *Brit Mus (Nat Hist) Entomology Dept*
'Novitates Staintonianae': illustrated catalogue of the 103 species of British lepidoptera added to his collection in 1842. *Hope Libr, University Mus, Oxford*
Another copy of 'Novitates Staintonianae'. *Private*
Fragment of a projected account of the British tortricidae *c*1880. *Horniman Mus and Libr, London*

[538] STEBBING, Revd Thomas Roscoe Rede (1835-1926)
FRS, naturalist

Original drawings, incl those for his *Report on the amphipoda collected by HMS Challenger during . . . 1873-1876. Brit Mus (Nat Hist) Zoology Dept*

[539] STEPHENSON, George (1781-1848)
Railway engineer

Report on the Shannon navigation 1831; miscellaneous papers 1816-45. *Science Mus, Newcastle upon Tyne* [NRA 9195]
Report on a proposed railway through Lord Carlisle's coalfields to Newcastle upon Tyne 1824. *Dept of Palaeography and Diplomatic, Durham Univ* (C 590/15) [NRA 11493]
Report on proposed railway communications with Ireland 1838. *Nat Libr of Wales* (MS 4880)
Miscellaneous correspondence and papers *c*1819-1848, incl specification rel to a locomotive steam engine patent 1846. *Inst of Mechanical Engineers* [NRA 9515]
Correspondence with E Pease rel to the Stockton and Darlington railway. *Darlington Public Libr*
Letters to WT Salvin 1826-41. *Durham County RO* (D/Sa/C 139) [NRA 10404]
Miscellaneous letters (14) 1824-29. *Liverpool RO*

See also Stephenson R

[540] STEPHENSON, Robert (1803-1859)
FRS, civil engineer

Drawings (1 vol) for bridges etc for the London & Birmingham Railway 1838-40. *Huntington Libr, San Marino*
Correspondence (1 vol) with E Clark rel to his bridges over the Conway and the Menai Straits 1846-47. *Inst of Civil Engineers*
Letters to his parents 1824-26 and rel to the Conway bridge 1848-49. *Nat Libr of Scotland*
Report on proposed improvements to Seaham Harbour 1856. *Durham County RO* (Londonderry Papers) [NRA 11528]
Correspondence with WT Salvin 1828-32. *Ibid* (D/Sa/C 143) [NRA 10404]
Correspondence with S Smiles about his biography of George Stephenson 1851-58. *Devon RO, Exeter* (Hartree Papers) [NRA 11697]
Correspondence (12 items) with Sir GB Airy 1850-52. *R Greenwich Observatory, Herstmonceux* (in MSS 943-48)
Correspondence (11 items) with JD Forbes 1848-56. *St Andrews Univ Libr* [NRA 13132]
A few legal and other papers 1827-58. *Inst of Mechanical Engineers* [NRA 9515]

[541] **STEVENSON, Robert** (1772-1850)
Civil engineer

Report on improving the navigation of the Forth 1829; travel journal 1817, personal and family correspondence. *Nat Libr of Scotland* (Acc 3056; MSS 2674, 3444, 3831-32, 3899, 3918)
Illustrated journal 1801-06. *Stevenson Coll of Further Education, Edinburgh* [NRA 23492]
Plan and sections for the improvement of Pulteney harbour, Wick 1822. *Private* [NRA 11847]
Papers and correspondence on the improvement of the Grassmarket, Edinburgh 1824. *Private* [NRA 13863]
Report and letters rel to the Fossdyke navigation 1840-44. *Lincolnshire Archives Office* [NRA 11494]

[542] **STEWART, Balfour** (1828-1887)
FRS, physicist and meteorologist

Reports, papers and correspondence, incl *c*280 letters to Sir E Sabine 1859-72. *Royal Soc*
Letters (162) to Sir GG Stokes 1860-87. *Cambridge Univ Libr* (in MS Add 7656)
Correspondence (68 items) with JD and G Forbes 1856-80. *St Andrews Univ Libr* [NRA 13132]
Letters (27) to officers of the Royal Astronomical Society 1864-87. *R Astronomical Soc*
Correspondence (13 items) with Sir GB Airy 1858-70. *R Greenwich Observatory, Herstmonceux* (in MSS 948-52)
Correspondence (13 items) with Lord Kelvin 1861-80. *Cambridge Univ Libr* (in MS Add 7342)
Correspondence with the Revd SJ Perry 1883-86. *Stonyhurst Coll*

See also De La Rue

[543] **STOKES, Sir George Gabriel,** 1st Bt (1819-1903)
FRS, mathematician and physicist

Correspondence (*c*20,000 items) and papers 1836-1903, incl notes of lectures attended and given, working papers and MSS of his published works. *Cambridge Univ Libr* (MS Add 7656)
Reports on papers submitted to the Royal Society 1851-90 and correspondence (*c*180 items); correspondence (68 items) with Sir JFW Herschel 1852-70. *Royal Soc*
MS rel to the refraction of light 1851. *Pembroke Coll, Cambridge*
Letters (120) to J Tyndall. *Royal Inst*
Letters (34) to Sir J Conroy 1875-84. *Balliol Coll, Oxford* [NRA 22851]
Letters (26) to W Sharpey 1860-66. *University Coll London* (MS Add 227) [NRA 18112]
Letters (18) to M Faraday 1852-60. *Inst of Electrical Engineers*
Correspondence (15 items) with Sir GB Airy 1849-66. *R Greenwich Observatory, Herstmonceux* (in MSS 941-51)
Correspondence (12 items) with Lord Kelvin 1859-98. *Glasgow Univ Libr*
Letters (10) to JC Maxwell 1857-79. *Cambridge Univ Libr* (in MS 7655) [NRA 24528]

▷ DB Wilson, *Catalogue of the manuscript collections of Sir George Gabriel Stokes and Sir William Thomson, Baron Kelvin of Largs, in Cambridge University Library*, 1976

See also Abel, Adams, Airy, Andrews T, Armstrong WG, Babbage, Ball RS, Barrett, Brewster, Brodie BC the younger, Burdon-Sanderson, Carpenter WB, Cayley A, Crookes, Darwin GH, De La Rue, De Morgan, Dewar, Fairbairn, Forbes G, Forbes JD, Foster M, Frankland, Galton, Gassiot, Gill, Hamilton DJ, Harcourt AGV, Harcourt WVV, Hartley WN, Herschel J, Herschel JFW, Hooker JD, Huggins, Huxley TH, Joule, Lamb, Lloyd H, Lubbock J, Marshall J, Matthey, Maxwell, Peacock, Pengelly, Phillips J, Pole, Powell B, Prestwich, Ramsay W, Robinson TR, Roscoe, Sabine, Schuster, Sharpey, Smithells, Smyth, Sorby, Spottiswoode, Stewart, Stoney, Strutt JW, Sylvester, Tait, Thomson JJ, Turner HH, Tyndall

[544] **STONER, Edmund Clifton** (1899-1968)
FRS, physicist

Correspondence 1909-66 and papers, incl his school and university notebooks 1909-24, notes on books read 1918-34, diaries 1919-68, administrative and committee papers 1940-63, MSS of his published and unpublished works, and autobiographical collections 1909-63. *Brotherton Libr, Leeds Univ* [NRA 17735]

See also Blackett, Bragg WL, Cockcroft, Hume-Rothery

[545] **STONEY, George Johnstone** (1826-1911)
FRS, physicist

Letters (69) to officers of the Royal Astronomical Society 1864-1900. *R Astronomical Soc*
Letters (19) to Sir GG Stokes 1867-91. *Cambridge Univ Libr* (in MS Add 7656)

[546] **STRATFORD, William Samuel** (1791-1853)
FRS, astronomer

Abstracts prepared for the *Monthly Notices* of the Royal Astronomical Society *c*1827-29, papers submitted to the Society, 28 letters to its officers 1832-52, and letters to J Lee. *R Astronomical Soc*
Papers, incl calculations rel to Halley's comet 1835-36. *Cambridge Univ Libr* (MSS Add 4236-38)
Letters (65) to Sir JW Lubbock 1829-47; correspondence (26 items) with Sir JFW Herschel 1826-47. *Royal Soc*
Correspondence (41 items) with Sir GB Airy 1835-52. *R Greenwich Observatory, Herstmonceux* (in MSS 938-44)
Letters (40) to Sir JFW Herschel 1844-47. *Texas Univ, Austin*

[547] **STRUTT, John William,**
3rd Baron Rayleigh, OM (1842-1919)
FRS, physicist

Experimental notebooks (11) 1862-1919; rough notes and calculations; MSS (187) of his published papers; lecture notes; papers and correspondence as chief scientific adviser to Trinity House; microfilm of his scientific correspondence. *Air Force Cambridge Research Laboratories, Bedford, Mass* [NRA 18694]
Scientific letters addressed to him (*c*1,100) 1871-1919; congratulatory letters addressed to him (200); correspondence (100 items) with Sir W Ramsay 1887-1907; calculations and data readings (*c*100 sheets). *Private*
Notebook; miscellaneous letters (12). *Royal Inst*
Comments, corrections and marginal notes to offprints (1860-70). *John Rylands Univ Libr, Manchester*
Correspondence (50 items) with Sir OJ Lodge 1885-1916; letters (27) to Sir W Ramsay 1886-1916. *University Coll London* [NRA 20647]
Letters (43) to Sir GG Stokes 1873-98. *Cambridge Univ Libr* (in MS Add 7656)
Letters (31) to Lord Kelvin 1867-1907. *Ibid* (in MS Add 7342)
Letters (28) to Macmillan & Co 1873-1918. *Brit Libr* (in Add MS 55220)
Letters (13) from him and his wife to O Browning 1879-*c*1887. *Hastings Public Libr*
Letters (10) to Sir JJ Thomson. *Cambridge Univ Libr* (in MS Add 7654)

See also Galton, Lamb, Larmor, Lodge, Ramsay W, Schuster, Strutt RJ, Tait

[548] **STRUTT, Robert John,**
4th Baron Rayleigh (1875-1947)
FRS, physicist

Experimental notebooks (23) *c*1904-47. *Air Force Cambridge Research Laboratories, Bedford, Mass* [NRA 18694]
Interleaved and annotated copy of *John William Strutt, 3rd Baron Rayleigh* (1924). *Private*
Correspondence (2 files) rel to college administration 1936-47. *Imperial Coll, London* [NRA 9526]
Correspondence (1 vol) with TDN Besterman on spiritualism 1932. *Brit Libr* (Add MS 57729)
Correspondence (12 items) with Sir OJ Lodge 1922-38. *Incorporated Soc for Psychical Research* [NRA 11857]
Miscellaneous correspondence 1906-32. *Inst of Geological Sciences* [NRA 18675]
Correspondence with GE Hale 1907-32. *California Inst of Technology, Pasadena*

[549] **STUART, Alexander** (1673-1742)
FRS, physician

Papers 1698-1725, incl prescriptions (1 vol) 1698-1713, medical and surgical observations and experiments (2 vols) made on voyages to the E Indies 1704-07, case notes (7 vols) 1708-25, notes of clinical lectures by H Boerhaave 1709-10, catalogue of his books and other papers rel to his studies at Leiden 1709-11, notes of his fees, and published and unpublished medical treatises. *Hunterian Libr, Glasgow Univ* (MSS 50, 225, 401, 525-38, 555, 569-72)
Medical observations (1 vol) with W Wasey at Westminster Infirmary 1723-24. *R Coll of Physicians, London*

[550] **STURGEON, William** (1783-1850)
Electrician

A few letters from him. *Manchester Literary and Philosophical Soc*

[551] **SWAINSON, William** (1789-1855)
FRS, naturalist

Papers rel to ornithology, entomology and zoology, MSS of his published works, and some correspondence. *Balfour and Newton Libr, Zoology Dept, Cambridge*
Scientific and family papers and correspondence 1833-48, with drawings of birds 1829. *Nat Libr of Australia, Canberra*
Journals of his travels in Italy 1810-13. *New Plymouth Mus, New Zealand*
Scientific papers, incl notes on the classification of birds and fishes, sketches and notes on eucalypts, and material for his published works. *Alexander Turnbull Libr, Wellington*

Letters (934) addressed to him 1806-40, with 5 drawings of New Zealand trees 1845-49. *Linnean Soc*

Notes on Australian trees 1853. *Mitchell Libr, Sydney*

Correspondence rel to zoological illustrations 1828-32. *Radcliffe Science Libr, Oxford*

Letters (27) to Sir WJ Hooker 1830-42, with 10 sketches of Australasian trees. *R Botanic Gardens, Kew*

Drawings (56, with 45 plate proofs) of birds, flora and scenery 1808-54. *Auckland Inst and Mus*

Drawings (1 vol) of birds (60), reptiles (21) and New Zealand scenery (6). *Nat Mus, Wellington*

Drawings (43) of New Zealand scenery. *Nat Art Gallery, Wellington*

▷ACLG Günther, 'An index of William Swainson's correspondence', *Proceedings of the Linnean Society*, xii, 1899-1900, pp14-61; NF McMillan, 'William Swainson FRS in New Zealand with notes on his drawings held in New Zealand', *J Soc Bibliog Nat Hist*, ix, pt 2, 1979, pp161-69

See also Burchell, Gray, Hooker WJ, Jardine, Westwood

[552] **SWAN, Sir Joseph Wilson** (1828-1914)
FRS, chemist and electrical inventor

Scientific, personal and family correspondence and papers, incl laboratory notebooks 1885-1914 and other working papers mainly rel to photography 1863-1911, electric lighting 1876-1905, and the gas cell 1903-14, business papers incl records of the Swan Electric Light Co 1881-96, drafts of lectures and speeches *c*1863-1914, papers rel to public affairs *c*1872-1914, autobiographical writings 1906-14, and 108 letters to Lady Swan 1881-1914. *Tyne and Wear Archives Dept, Newcastle upon Tyne* [NRA 24634]

[553] **SWYNNERTON, Charles Francis Massy** (1877-1938)
Entomologist

Entomological notes. *Hope Libr, University Mus, Oxford*

[554] **SYDENHAM, Thomas** (1624-1689)
Physician

Observations on fevers and acute diseases 1669-*c*1680; treatise on the gout 1683. *R Coll of Physicians, London*

'Anatomie' 1668; 'De arte medica' 1669; 'Smallpox' 1669; 'Tussis' nd. *Public Record Office* (PRO 30/24/47/2)

'De variolis' 1669; 'A dysentery' 1670; 'Febres intercurrentes' 1670; 'Variola' 1670; 'Pleurisie' 1670; 'Febres intermittentes' 1670; 'Of the four constitutions' 1675; 'Tussis' nd; letters (5) to J Locke 1767-79. *Bodleian Libr, Oxford* (in MSS Locke c 19, 29, 42, f 21) [NRA 16992]

'Theologia rationalis'. *Cambridge Univ Libr* (MS Dd iii 75)

Notebook containing 17th century transcripts of his writings. *Bodleian Libr, Oxford* (MS Rawlinson c 406)

▷K Dewhurst, *Dr Thomas Sydenham (1624-1689): his life and original writings*, 1966

[555] **SYKES, William Henry** (1790-1872)
FRS, naturalist

First and second reports (11 vols) on statistics of the Deccan 1826, 1829. *India Office Libr and Records* (MSS Eur D 140-50)

Notes and sketches (10 vols) on the economic plants and agriculture of the Deccan 1824-31. *Brit Mus (Nat Hist) General Libr*

Paper submitted to the Royal Geographical Society on tribes in Travancore 1859. *R Geographical Soc*

Correspondence (36 items) with and rel to the Marquess of Dalhousie 1847-56. *Scottish Record Office* (GD 45/6/133, 14/611) [NRA 17164]

Correspondence (21 items) with Sir JFW Herschel 1839-62. *Royal Soc*

Letters (15) to Sir WJ Hooker 1842-55. *R Botanic Gardens, Kew*

[556] **SYLVESTER, James Joseph** (1814-1897)
FRS, mathematician

Correspondence (900 items) and papers, incl mathematical notes, lecture notes and poems. *St John's Coll, Cambridge* [NRA 9502]

Letters from him 1837-96, incl 56 to TA Hirst 1859-85, 41 to Lord Brougham 1837-63, and 10 to Countess Edith Gigliucci 1865-96. *University Coll London* (in London Mathematical Society Papers; Brougham Papers; MS Add 221; College Corresp)

Miscellaneous letters (107) 1842-97. *Brown Univ Libr, Providence, Rhode Island*

Letters (60) to WJC Miller and others 1856-93. *Columbia Univ Libraries, New York*

Letters (19) to Sir AB Kempe 1875-96. *West Sussex RO, Chichester* [NRA 17595]

Letters (18) to S Newcomb 1877-91. *Libr of Congress, Washington*

Letters (16) to Sir JN Lockyer. *Exeter Univ Libr*

Letters (14) to MG Mittag-Leffler. *Mittag-Leffler Inst, Djursholm, Sweden*

Letters (12) to Sir GG Stokes 1863-88. *Cambridge Univ Libr* (in MS Add 7656)
Letters (12) to J Tyndall. *Royal Inst*
Letters (11) to Lord Kelvin 1845-92. *Cambridge Univ Libr* (in MS Add 7342)
Letters (10) to C Babbage 1835-69. *Brit Libr* (Add MSS 37189-99 *passim*)

▷RC Archibald, 'Unpublished letters of James Joseph Sylvester and other new information concerning his life and work', *Osiris*, i, 1936, pp85-154

See also Cayley A

[557] **TAIT, Peter Guthrie** (1831-1901)
Mathematician and physicist

Lectures 1881-82; correspondence 1869-91. *Edinburgh Univ Libr* (MSS Dc 2.76, 5.98-99)
MS of his *Treatise on the dynamics of a particle* (1856). *Peterhouse, Cambridge*
Correspondence (169 items) with Lord Kelvin 1861-1901. *Cambridge Univ Libr* (in MS Add 7342)
Correspondence (167 items) with JC Maxwell 1859-79. *Ibid* (in MS 7655) [NRA 24528]
Correspondence (151 items) with Sir GG Stokes 1860-99. *Ibid* (in MS 7656)
Correspondence (103 items) with Lord Kelvin 1861-98. *Glasgow Univ Libr*
Letters (*c*100) to Sir WR Hamilton 1858-65. *Trinity Coll, Dublin* [NRA 20076]
Letters (16) to Lord Rayleigh 1891-1902. *Private*
Letters (13) to JD Forbes 1860-67. *St Andrews Univ Libr* [NRA 13132]

[558] **TALBOT, William Henry Fox** (1800-1877)
FRS, pioneer of photography

Correspondence and papers rel to photography, botany, biblical archaeology, and business and family affairs, incl scientific notebooks (2) 1834-38, and photographic correspondence (3 box files) 1831-77. *Fox Talbot Mus, Lacock, Wiltshire*
Correspondence and papers 1837-76, incl notebook 1839-40, and 88 letters from Sir D Brewster 1839-65. *Science Mus Libr*
Annotated copy of D Turner and LW Dillwyn, *The botanist's guide through England and Wales* (1805). *Nat Libr of Wales*
Correspondence (92 items) with Sir JFW Herschel 1826-51; letters (28) to Sir JW Lubbock 1834-41. *Royal Soc*
Letters (42) to Sir WJ Hooker 1831-59. *R Botanic Gardens, Kew*
Letters (15) to C Babbage 1831-44. *Brit Libr* (Add MSS 37186-37201 *passim*)
Letters (24) to Sir WC Trevelyan. *Newcastle Univ Libr* [NRA 12238]
Letters (13) to Sir WR Grove. *Royal Inst*
Correspondence (11 items) with JD Forbes 1839-67. *St Andrews Univ Libr* [NRA 13132]

See also Brewster

[559] **TAYLOR, Brook** (1685-1731)
FRS, mathematician

Letters (64) as secretary of the Royal Society. *Royal Soc*

[560] **TAYLOR, Sir Geoffrey Ingram,** OM (1886-1975)
FRS, aerodynamicist

Scientific, personal and family papers and correspondence, incl notebooks, notes, patents, reports, committee papers, articles and addresses. *Trinity Coll, Cambridge* [NRA 22889]
Correspondence with Viscount Cherwell 1916-51. *Nuffield Coll, Oxford* [NRA 16447]

See also Adrian

[561] **TAYLOR, John William** (1845-1931)
Conchologist

Notebooks (208) of data (partly unpublished) for his *Monograph of the land and freshwater mollusca of the British Isles* (1894-1921); annotated volumes (2) of the *Monograph*; notes and correspondence; list of his shell collection. *Brit Mus (Nat Hist) Zoology Dept*
Lecture notes, working papers, drawings, correspondence and papers rel to his *Monograph*, and record compiled with W Nelson of the mollusca of the West Riding 1891. *Leeds City Archives Dept*
MS (incomplete) of his *Monograph*, with original drawings. *Leeds City Mus*
Annotated published floras, and correspondence with Elsie and KM Morehouse. *Doncaster Mus and Art Gallery*

[562] **TEBBUTT, John** (1834-1916)
Astronomer and meteorologist

Astronomical and meteorological papers 1853-1910, diary 1859-63, and correspondence (46 vols) 1847-1915. *Mitchell Libr, Sydney*
Papers (2) submitted to the Royal Astronomical Society 1878, and letters (52) to its officers 1868-1900. *R Astronomical Soc*

[563] **TELFORD, Thomas** (1757-1834)
FRS, engineer

Notebooks (8) of architectural memoranda and work in progress; drawings (2 vols) of bridges, canals and railways 1810-30; correspondence and reports (2 vols) on Dean Bridge, Edinburgh 1828-32;

correspondence (*c*1,100 items) and papers, mainly rel to the Gotha canal, Sweden, Holyhead and S Wales road surveys, London Bridge, and Trent and Mersey navigation. *Inst of Civil Engineers* [NRA 14021]
Papers, incl two drafts of his autobiography, reports, notes and correspondence rel to the Gloucester and Berkeley canal (over 300 items), and other projects in England and Wales. *Ironbridge Gorge Mus, Telford, Shropshire* [NRA 22554]
Letter books 1790-94 (4 vols) and other papers rel to the British Fisheries Society; Caledonian canal letter books 1803-33 (9 vols); reports, plans and papers on Highland roads and bridges *c*1779-1834. *Scottish Record Office* [NRA 13863]
Notebook 1833; miscellaneous correspondence and accounts rel to various projects in Scotland and England *c*1800-1834. *Nat Libr of Scotland*
Reports, surveys, estimates, specifications and correspondence (over 1,000 items) rel to his work for the Commissioners for Highland Roads and Bridges and the Caledonian canal 1803-34. *House of Lords RO*
Reports, accounts, correspondence and papers concerning numerous road, bridge and canal projects in England and Wales. *Public Record Office* (British Transport Historical Records and records of the Ministries of Transport and Works)
Reports, estimates, plans and correspondence as county surveyor. *Shropshire RO, Shrewsbury*
Plans, estimates, accounts and correspondence rel to Bewdley bridge 1795-1804. *Hereford and Worcester RO, Worcester* [NRA 12165]
Aberdeen harbour plans 1802-31. *Aberdeen District Archives*
Reports and correspondence mainly rel to the Clyde docks and the Forth and Clyde canal 1806-26. *Strathclyde Regional Archives, Glasgow* [NRA 13871]
Drawings etc rel to roads connecting England, Ireland and Scotland 1808. *Ewart Public Libr, Dumfries*
Plans (1 vol) for his report on the same 1809. *Cumbria RO, Carlisle*
Reports and correspondence rel to Morpeth bridge 1810-30, with other papers. *Northumberland RO, Newcastle upon Tyne*
Dundee harbour plans (33) 1815-31. *Dundee District Archive and Record Centre* [NRA 13496]
Reports on the river Dee navigation 1817-29. *Clwyd RO, Hawarden*
Reports to the St Albans turnpike trust 1817-26. *Hertfordshire RO*
Report on the Knaresborough canal 1818. *Yorkshire Archaeological Soc, Leeds* (Slingsby Papers) [NRA 12891]
Report on proposed navigation to Norwich 1822. *Norfolk RO*
Reports on Boston haven 1823. *Boston Reference Libr, Lincs*
Report on King's Lynn harbour and other papers. *Nottingham Univ Libr*
Papers, incl letters, copies of reports, designs for a new London Bridge, and a plan of Thames borings 1823. *Corporation of London RO*
Reports on the rebuilding of Brentford bridge 1823-24; letters and reports rel to rebuilding part of the river wall 1826-27. *Greater London RO*
Report on Seaham harbour and correspondence with J Buddle 1823-33. *Durham County RO* (Londonderry Papers) [NRA 11528]
Reports, correspondence and plans rel to bridges at Over and Stone, S Wales turnpike roads and other work for the county 1824-34. *Gloucestershire RO* [NRA 9191]
Correspondence, papers and a plan rel to Somerset turnpike roads 1825-26. *Somerset RO, Taunton*
Proposed improvements to Folkestone harbour 1829. *Kent Archives Office, Folkestone*
Report on Dover harbour 1834; plan of Loose viaduct. *Ibid, Maidstone*
Correspondence with A Little 1780-1803. *Private*
Correspondence rel to steam engines. *Birmingham Reference Libr* (Boulton & Watt Colln)
Account with C Hoare & Co, bankers, 1802-25. *Private*

▷AE Penfold, 'A guide to sources for a study of the life and work of Thomas Telford', *Business Archives*, xliii, 1977, pp7-29

See also Barlow

[564] **THISELTON-DYER, Sir William Turner,** KCMG (1843-1928)
FRS, botanist

Correspondence and papers, incl diaries (2 vols) to 1926, notes for lectures 1871-92, notes (2 vols) on Greek and Latin plant names 1911-*c*1926, letters and papers (1 vol) on the Kelvin-Dyer controversy 1903, letters from CR Darwin 1873-81, and Sir JD Hooker 1870-1909. *R Botanic Gardens, Kew*
Miscellaneous correspondence, incl 18 letters from GJ Romanes 1882-93. *Linnean Soc*
Correspondence (14 items) with TH Huxley 1875-94. *Imperial Coll, London*

[565] **THOMPSON, Sir Benjamin,** Count von Rumford (1753-1814)
FRS, man of science

Correspondence and papers (2 boxes). *Houghton Libr, Harvard Univ, Cambridge, Mass*

Correspondence and papers (627 items) of Thompson and his daughter from 1755. *American Academy of Arts and Sciences, Boston, Mass*

Travel journal, London to Munich 1801 (copy made by Lady Palmerston 1802). *Birmingham Univ Libr* (MSS 6/iv/29) [NRA 15237]

Miscellaneous correspondence and papers, incl notebook 1769-72. *New Hampshire Hist Soc, Concord*

Papers (15) submitted to the Royal Society. *Royal Soc*

Letters (65) to Lady Palmerston 1793-1804. *Dartmouth College, Hanover, New Hampshire* [NRA 10398]

Letters (*c*60) to W Savage and others *c*1799-1813. *Royal Inst*

Letters (48) to Sir J Banks 1784-1804. *Brit Libr* (in Add MSS 8096-99)

Correspondence (13 items) with his publishers Cadell & Davies 1799-1802. *Ibid* (in Add MS 34045)

Designs (27) for boilers, stoves and other kitchen equipment 1794-1802. *Royal Inst of British Architects*

▷ SC Brown, *Benjamin Thompson, Count Rumford*, 1979

[566] **THOMPSON, Sir D'Arcy Wentworth** (1860-1948)
FRS, zoologist

Correspondence (*c*30,000 items) 1878-1948 and papers, incl scientific notebooks, working papers, lecture notes, drawings, MSS of his published works, and papers rel to international fishery research and the International Commission for the Exploration of the Sea. *St Andrews Univ Libr*

Miscellaneous notes, testimonials, photographs and some correspondence 1884-1946. *Dundee Univ Libr* (Peacock Papers) [NRA 19242]

Letters to him from naturalists 1934-47. *R Scottish Mus, Edinburgh*

[567] **THOMPSON, Silvanus Phillips** (1851-1916)
FRS, physicist

Correspondence (*c*600 items) 1876-1916 and papers, incl MS (4 vols) of his *William Thomson, Baron Kelvin of Largs* (1910) and a lecture on the compass 1907. *Imperial Coll, London* [NRA 11421]

MSS on permanent magnets and harmonic analysis, miscellaneous correspondence, papers, sketches and photographs, together with his own books, pamphlets and collected MSS. *Inst of Electrical Engineers* [NRA 20573]

Correspondence (92 items) with Sir OJ Lodge 1883-1916. *University Coll London* [NRA 20647]

Correspondence (19 items) with Lord Kelvin 1884-1905. *Cambridge Univ Libr* (in MS Add 7342)

Letters (13) to officers of the Royal Astronomical Society 1875-96. *R Astronomical Soc*

Miscellaneous correspondence (12 items). *Sussex Univ Libr, Brighton*

See also Crookes, Lodge, Thomson W

[568] **THOMSON, Allen** (1809-1884)
FRS, biologist

Journal (6 vols) of a continental tour 1833, lecture notes 1844-*c*1877, paper on the formation of the egg and evolution of the chick 1829-30, and letters addressed to him. *R Coll of Surgeons, London*

Papers and correspondence. *Nat Libr of Scotland* (in MSS 9235-37)

[569] **THOMSON, Sir Charles Wyville** (1830-1882)
FRS, naturalist

Papers rel to the *Challenger* expedition commission; papers (2) submitted to the Royal Society; miscellaneous correspondence (24 items). *Royal Soc*

Letters (71) from AE Agassiz 1867-81; lectures 1848-49; student notes of his lectures 1872. *Edinburgh Univ Libr* (MS Da; MS Dk 6.13; MS Ewa 7)

Paper on zoophytes recorded in the north of Ireland *c*1857. *Ulster Mus, Belfast*

Letters (31) to Macmillan & Co 1872-80. *Brit Libr* (in Add MS 55218)

Correspondence (20 items) with TH Huxley *c*1857-80. *Imperial Coll, London*

[570] **THOMSON, Sir George Paget** (1892-1975)
FRS, physicist

Scientific correspondence and papers, incl notes on electron diffraction and thermonuclear reaction, notes and drafts for lectures and publications and draft autobiography. *Trinity Coll, Cambridge*

Maud Committee papers (1 box) rel to atomic energy 1939-42. *Churchill Coll, Cambridge*

Correspondence (28 items) with N Bohr 1930-56. *Niels Bohr Inst, Copenhagen*

Correspondence with Viscount Cherwell 1917-56. *Nuffield Coll, Oxford* [NRA 16447]

Correspondence with W Cochrane 1934-61. *Glasgow Univ Archives*

[571] **THOMSON, John** (1765-1846)
FRS, physician and surgeon

Correspondence and papers. *Nat Libr of Scotland* (in MSS 9235-37)
Prescription book *c*1815, and student notes (4 vols) of his lectures on surgery 1808-11. *R Coll of Physicians, Edinburgh* [NRA 16015]

[572] **THOMSON, Sir Joseph John,** OM (1856-1940)
FRS, physicist

Correspondence (over 4,000 items) and papers, incl *c*400 laboratory notebooks, MSS of his published works and working papers. *Cambridge Univ Libr* (MS Add 7654)
Referee reports (35) 1885-99, and *c*130 letters from him 1884-1934, incl 25 to Sir A Schuster. *Royal Soc*
Miscellaneous correspondence and papers, incl notes and drafts for his published works. *Trinity Coll, Cambridge* [NRA 23828]
Correspondence (41 items) with Lord Rutherford 1895-1935. *Cambridge Univ Libr* (in MS Add 7653)
Letters (20) to Sir OW Richardson. *Texas Univ, Austin* [NRA 22337]
Correspondence (13 items) with Lord Kelvin 1882-1906. *Cambridge Univ Libr* (in MS Add 7342)
Letters (10) to Sir GG Stokes 1882-1901. *Ibid* (in MS Add 7656)
Correspondence with GE Hale 1912-21. *California Inst of Technology, Pasadena*

▷ *Notes and Records*, x, 1953, contains an index to the correspondence now in Cambridge University Library

See also Dewar, Lodge, Schuster, Strutt JW, Threlfall

[573] **THOMSON, William,**
Baron Kelvin of Largs, OM, GCVO (1824-1907)
FRS, physicist

Personal and scientific correspondence and papers, incl *c*5,600 letters addressed to him, 31 letter books 1878-84, 1889-1906 containing drafts of *c*4,000 out-letters, 180 research or other notebooks and travel journals 1835-1907, and drafts of his publications. *Cambridge Univ Libr* (MS Add 7342)
Business and scientific correspondence and papers, incl laboratory records 1870-98, reports on ships' trials 1872, drafts and proofs of his publications, 5 letter books 1883-99; business records of Kelvin & White Ltd, Kelvin & Hughes Ltd, and Kelvin, Bottomley & Baird Ltd from 1876. *Glasgow Univ Libr* and *Glasgow Univ Archives* [NRA 13691]
Letters and reports on a device for taking soundings through a ship's hull 1905-07. *Nat Maritime Mus, Greenwich*
Letters (55) to JD Forbes 1846-65. *St Andrews Univ Libr* [NRA 13132]
Letters (43) to S de Ferranti 1882-84. *Private* [NRA 13215]
Letters (40) to Sir OJ Lodge 1884-1907. *University Coll London* (in MS Add 89)
Miscellaneous correspondence (37 items). *Royal Inst*
Letters (21) to M Faraday 1845-60. *Inst of Electrical Engineers* (Tyndall Colln)
Letters (19) to SP Thompson 1882-1904. *Imperial Coll, London* [NRA 11421]
Correspondence (12 items) with JC Maxwell 1868-79, and comments on papers by Maxwell. *Cambridge Univ Libr* (in MS Add 7655)
Letters (12) to Sir A Schuster 1882-1904. *Royal Soc*

▷ DB Wilson, *Catalogue of the manuscript collections of Sir George Gabriel Stokes and Sir William Thomson, Baron Kelvin of Largs, in Cambridge University Library*, 1976, and *Kelvin Papers: index to the manuscript collection of William Thomson, Baron Kelvin in Glasgow University Library*, 1977

See also Andrews T, Boole, Cayley A, Darwin GH, De Morgan, Dewar, Ellis RL, Ewing, Ferranti, Fitzgerald, Forbes JD, Gill, Glazebrook, Hamilton WR, Joule, Lamb, Larmor, Lodge, Maxwell, Ramsay W, Routh, Sabine, Schuster, Stewart, Stokes, Strutt JW, Sylvester, Tait, Thiselton-Dyer, Thompson SP, Thomson JJ, Townsend, Tyndall

[574] **THORPE, Sir Jocelyn Field** (1872-1940)
FRS, chemist

Papers on the scope of organic chemistry, research and scientific industrial development; notes (3 vols) for lectures on organic chemistry 1914-15; administrative correspondence (2 files) 1915-32. *Imperial Coll, London*

[575] **THORPE, Sir Thomas Edward** (1845-1925)
FRS, chemist

Papers and letters (127 items). *Royal Soc*
Tables of experimental data *c*1872-76; correspondence (30 items) 1875-1924, incl 21 letters to A Smithells 1877-1924. *Brotherton Libr, Leeds Univ* (MSS 346, 391, 416) [NRA 20283]
Administrative correspondence and papers 1909-21. *Imperial Coll, London* [NRA 9526]

[576] **THRELFALL, Sir Richard,** GBE (1861-1932)
FRS, physicist and chemical engineer

Correspondence and papers, incl notes and reports. *Birmingham Reference Libr*

Correspondence and papers *c*1887-1904, incl letter books, letters addressed to him and testimonials. *Sydney Univ Archives* (P 28)

Correspondence (42 items) with Sir JJ Thomson 1886-96. *Cambridge Univ Libr* (in MS Add 7654)

[577] **THUDICHUM, John Louis William** (1829-1901)
Physician and chemist

Notebooks 1858-1901. *Nat Inst for Medical Research, Medical Research Council*

Miscellaneous papers. *Royal Soc*

[578] **THWAITES, George Henry Kendrick** (1811-1882)
FRS, botanist and entomologist

Correspondence 1852-79, incl 198 letters to Sir WJ Hooker 1846-65. *R Botanic Gardens, Kew*

Letters (24ff) to W Wilson 1843-46. *New York Botanical Garden Libr*

[579] **TIZARD, Sir Henry Thomas,** GCB (1885-1959)
FRS, chemist

Correspondence and papers (714 files) 1891-1959, incl diaries and diary notes 1927, 1937-45, appointment diaries and notebooks 1931-57, autobiography to 1929, MSS of his published and unpublished works, lectures, broadcasts and speeches 1906-59, papers rel to his missions to the USA 1940 and Australia 1943, as scientific adviser to the Chief of Air Staff 1939-43, and at the Ministry of Aircraft Production 1940-43. *Imperial War Mus* [NRA 24549]

Correspondence as scientific adviser to the Chief of Air Staff 1939-42, and administrative correspondence and papers 1929-42. *Imperial Coll, London*

Correspondence with Viscount Cherwell 1913-50. *Nuffield Coll, Oxford* [NRA 16447]

Correspondence with Lord Blackett 1938-58. *Royal Soc* [NRA 22627]

Letters (11) to CT Onions 1942-47. *Birmingham Univ Libr* [NRA 14330]

[580] **TODD, Sir Charles,** KCMG (1826-1910)
FRS, astronomer

Diaries, notebooks, reports and correspondence 1852-1901. *South Australian Archives, Adelaide*

Letters (18) to officers of the Royal Astronomical Society 1861-99. *R Astronomical Soc*

Letters to HYL Brown *c*1893-99. *Geological Survey of South Australia, Adelaide*

[581] **TOWNSEND, Sir John Sealy Edward** (1868-1957)
FRS, physicist

Papers, incl drafts of his published works *c*1943-1956, patents and contracts 1914-25. *Bodleian Libr, Oxford* [NRA 17594]

Correspondence with Viscount Cherwell 1932-*c*1957. *Nuffield Coll, Oxford* [NRA 16447]

Letters (19) to Lord Rutherford 1898-1915. *Cambridge Univ Libr* (in MS Add 7653)

Correspondence (11 items) with Lord Kelvin 1899-1901. *Ibid* (in MS Add 7342)

[582] **TRAVERS, Morris William** (1872-1961)
FRS, chemist

Correspondence and papers 1889-1961, incl diaries, autobiographical papers, annotated reports etc rel to his work in India, and papers and correspondence rel to his *Life of Sir William Ramsay* (1956). *University Coll London* [NRA 14260]

[583] **TREVELYAN, Sir Walter Calverley,** 6th Bt (1797-1879)
Naturalist

Correspondence and papers, incl journals and diaries 1812-78, account books 1815-52, and notes, sketches and papers rel to geology, natural history, phrenology, family history and Northumberland affairs. *Newcastle Univ Libr* [NRA 12238]

Miscellaneous correspondence and papers, incl antiquarian papers and letters from scientists. *Brit Libr* (in Add MSS 22289, 24965, 27409, 28872, 29708, 29718, 31026-27)

Letters (57) to Sir WJ Hooker 1831-65. *R Botanic Gardens, Kew*

See also Acland, Brewster D, Buckland, Daubeny, Forbes E, Gray, Greville, Hooker WJ, Lindley, Rolleston, Talbot, Watson, Winch NJ

[584] **TRIMEN, Roland** (1840-1916)
FRS, entomologist

Correspondence and papers 1858-1912, incl entomological and other journals 1858-87, notes and papers rel to S African butterflies and other lepidoptera, and copies of his published works. *R Entomological Soc*

Entomological correspondence. *Hope Libr, University Mus, Oxford*

[585] **TURNBULL, Herbert Westren** (1885-1961)
FRS, mathematician

Notes for and drafts of his published works. *Cambridge Univ Libr*

[586] **TURNER, Dawson** (1775-1858)
FRS, botanist

Papers, incl collections, drawings and catalogues mainly rel to Norfolk, extracts from rare books and MSS, and rough draft (6 vols) of his library catalogue 1839. *Brit Libr* (Add MSS 23013-67, 23106-07, 27967, 28652, 28655-57, 29738, 50484-89)
Annotated copies of his own publications. *Ibid* (7032.i.30; 8255.bb.78; C.45.b.4-5)
Correspondence (82 vols) 1790-1850. *Trinity Coll, Cambridge* (MSS O.13.1-32, O.14.1-50)
Memoranda of his correspondence with 115 botanists (2 vols); correspondence with Sir WJ Hooker 1805-50; 51 letters to Ellen Hutchins 1807-20; letters to W Borrer 1807-17; translation of F Weber and DMH Mohr, *Naturhistorische Reise durch einen Theil Schwedens* (1804). *R Botanic Gardens, Kew*
'Miscellanea Curiosa': 6 vols of his literary and other collections, partly in transcript, incl his own botanical notes on Manchester, 1549-1834. *Virginia Hist Soc, Richmond*
Journals of a Cornish tour 1799 and a French tour 1814; correspondence with SP Woodward. *Castle Mus, Norwich*
Annotated copy of his *Synopsis of the British fuci*, I (1802). *Brit Mus (Nat Hist) Botany Dept*
Correspondence and papers, incl catalogue of his tracts and pamphlets 1832 and correspondence on business and family affairs 1852-55, with related papers. *Norfolk RO* [NRA 17246]
Correspondence and papers, incl library catalogue 1839, correspondence with Sir Francis and Lady Palgrave 1812-52, and 40 letters from JS Cotman. *Private*
Letters (11) to I D'Israeli. *Bodleian Libr, Oxford* [NRA 0842]

▷WR Dawson, 'Sir Joseph Hooker and Dawson Turner', *J Soc Bibliog Nat Hist*, ii, pt 6, 1950, pp218-22, and 'Dawson Turner, FRS (1775-1858)', ibid, iii, pt 6, 1958, pp303-10; ANL Munby, *Cult of the autograph letter in England*, 1962, pp33-60

See also Banks, Talbot

[587] **TURNER, Herbert Hall** (1861-1930)
FRS, astronomer

Letters addressed to him 1902-03, and papers incl notes on determining the time of sunset *c*1902-04. *History of Science Mus, Oxford* (MSS Mus 39, 85, 156)
Letters (519) to officers of the Royal Astronomical Society 1885-1900, with 29 to its Joint Permanent Eclipse Committee 1902-12, and 26 concerning expeditions to observe the eclipse of 1889. *R Astronomical Soc*
Letters (39) to Sir D Gill 1903-10. *R Geographical Soc*
Letters (34) to K Pearson, with notes on papers by O Stumpe, Sir FW Dyson and WE Thackeray 1888-1909. *University Coll London*
Letters (16ff) to Gilbert Murray 1909-22. *Bodleian Libr, Oxford* [NRA 16865]
Letters (10) to Sir GG Stokes 1887-1900. *Cambridge Univ Libr* (in MS Add 7656)

[588] **TUTTON, Alfred Edwin Howard** (1864-1938)
FRS, crystallographer

MS (incomplete) of his unpublished work on National Standards. *Royal Soc*
Student notes on his lectures 1894-95; correspondence as a member of the Technical Optics Committee 1918. *Imperial Coll, London*
Correspondence (73ff) with Macmillan & Co 1909-36. *Brit Libr* (in Add MS 55224)

[589] **TYNDALL, John** (1820-1893)
FRS, natural philosopher

Notebooks, lecture notes, diary of experiments, journal, correspondence and press cuttings. *Royal Inst* [NRA 9522]
Drafts of literary works and correspondence 1861-92. *Brit Libr* (Add MS 53715)
Correspondence (253 items) with TH Huxley 1851-94. *Imperial Coll, London*
Correspondence (80 items) with Sir GG Stokes 1852-*c*1889. *Cambridge Univ Libr* (in MS Add 7656)
Correspondence (58 items) with Sir JFW Herschel 1851-70. *Royal Soc*
Correspondence (*c*40 items) about publications 1867-92. *Private*
Letters to Sir JD Hooker 1856-93. *R Botanic Gardens, Kew*
Letters (12) to Sir F Galton. *University Coll London*
Correspondence (10 items) with Lord Kelvin 1850-92. *Cambridge Univ Libr* (in MS Add 7342)

▷JR Friday, RM MacLeod and P Shepherd, *John Tyndall, natural philosopher, 1820-1893: catalogue of correspondence, journals and collected papers*, 1974

See also Airy, Darwin CR, Herschel JFW, Hooker JD, Huxley TH, Lyell, Sabine, Stokes, Sylvester

[590] **UNWIN, William Cawthorne** (1838-1933)
FRS, engineer

Correspondence (241 items) 1856-1931, and papers, incl research reports 1859-1919, notebooks (2) 1882-1905, memorandum on his career 1922, speeches (8) 1902-21, and drawings 1875-1906. *Imperial Coll, London* [NRA 10964]

See also Fairbairn

[591] **URE, Andrew** (1778-1857)
FRS, chemist

Letters addressed to him. *Andersonian Libr, Strathclyde Univ, Glasgow*

Paper on disinfecting the cargo and crew of a ship *c*1831. *R Coll of Physicians, London*

Letters to Sir T Makdougall Brisbane 1810-11. *Private* [NRA 11854]

[592] **VIGNOLES, Charles Blacker** (1793-1875)
FRS, civil engineer

Journals (13) 1824-62, with some accounts 1815 and correspondence 1836-38. *Brit Libr* (Add MSS 34528-36, 35071, 58203-06)

Correspondence (*c*1,400 items) and related papers *c*1813-1873. *Portsmouth City RO*

[593] **VINCE, Revd Samuel** (1749-1821)
FRS, mathematician and astronomer

Mathematical and astronomical papers; miscellaneous letters (7). *R Greenwich Observatory, Herstmonceux* (MSS 288-91) [NRA 22822]

Papers (14) submitted to the Royal Society. *Royal Soc*

Correspondence (18 items) with Sir W Herschel 1783-1806. *R Astronomical Soc*

[594] **VINES, Sydney Howard** (1849-1934)
FRS, botanist

Miscellaneous papers, incl annotated copy of his *Lectures on the physiology of plants* (1886); letters (7) from Sir JD Hooker 1892. *Bodleian Libr, Oxford* (MSS Sherard 7, 401-06) [NRA 6305]

[595] **WAGER, Lawrence Rickard** (1904-1965)
FRS, geologist

Correspondence 1924-65 and papers 1919-65, incl diaries 1919-60, laboratory and field notebooks and working papers 1920-59, lecture scripts and notes 1925-63, draft essays *c*1930-1965, and committee papers. *Royal Soc* [NRA 13901]

WALKER ARNOTT, see Arnott

[596] **WALLACE, Alfred Russel,** OM (1823-1913)
FRS, naturalist

Correspondence and papers 1856-1912, incl drafts of his published works (20 vols) 1885-1910, and correspondence (9 vols) on spiritualism, land nationalisation, socialism and anti-vaccination 1856-1912. *Brit Libr* (in Add MSS 38794, 46414-42)

Correspondence 1873-1913 and papers, incl Malayan journal (4 vols) 1856-61, American journal 1886-87, illustrated notes on Malayan butterflies (1 vol) 1854-62, other natural history notes (2 vols) 1854-59, list of Australian birds, drawings of Amazon palms, and annotated copies of published works. *Linnean Soc*

Descriptions of fishes etc of the Amazon and Rio Negro, with drawings (4 vols) of fishes of the latter 1850-52; notebooks (2) on birds of the Malay archipelago 1855-61. *Brit Mus (Nat Hist) Zoology Dept*

MSS of his published communications to the Zoological Society. *Zoological Soc of London*

Correspondence, and annotated publications about spiritualism. *Hope Libr, University Mus, Oxford*

Correspondence (69 items) with Macmillan & Co 1875-1913. *Brit Libr* (in Add MS 55221)

Correspondence with MB Slater 1886-1909. *Manchester Central Libr* (MS f 925 Wal)

Letters (24) to officers of the Royal Geographical Society 1853-1908. *R Geographical Soc*

Letters (13) to Sir F Galton 1869-1904. *University Coll London*

See also Darwin CR

[597] **WALLICH, George Charles** (1815-1899)
Naturalist

Notebooks (7) mainly rel to marine biology and his claims to discoveries 1860-98; annotated copies (2 vols) of articles by himself and others 1842-80. *Wellcome Hist Medical Libr* (MSS 4962-70)

Notebooks and annotated copies of his published works 1857-84. *Univ of British Columbia Libr, Vancouver*

Comments (1 vol) on the reports of the *Challenger* expedition; charts (2) of voyages between Portsmouth and Calcutta 1850-51, 1857; drawings (59) of microscopic marine animals; annotated copies of his published works. *Brit Mus (Nat Hist) Zoology Dept*

Notes and drawings of dredgings off Greenland from HMS *Bulldog* 1860. *Ibid, General Libr*

Notes and drawings of desmideae and diatomaceae. *Ibid, Botany Dept*

[598] **WALLICH, Nathaniel** (1786-1854)
FRS, botanist

Memoir on the mimoseae in M Vahl's herbarium, begun *c*1803; letters (1 vol) from W Roscoe 1813-30; correspondence (2 vols) 1841-52; letters (612) to Sir WJ Hooker 1818-54. *R Botanic Gardens, Kew*

Letters (17) to Sir J Banks 1818-20; notes kept by T Hardwicke (1 vol) of letters from Wallich 1820-25; copies (26 items) of his official correspondence 1822-25. *Brit Libr* (Add MSS 9869-70, 33982)

Botanical catalogues and other papers rel to Burma 1826-28. *Linnean Soc*

See also Clarke CB

[599] **WALLIS, Sir Barnes Neville** (1887-1979)
FRS, aeronautical engineer

Scientific papers and correspondence 1905-76, incl reports, working papers, technical drawings and photographs, relating particularly to airship design 1917-29 and weapons and unmanned or supersonic aircraft 1939-65. *Science Mus Libr*

Aeronautical research papers (2 boxes) 1940-58. *Churchill Coll, Cambridge* [NRA 22889]

Correspondence with Viscount Cherwell 1943-44. *Nuffield Coll, Oxford* [NRA 16447]

[600] **WALLIS, Revd John** (1616-1703)
FRS, mathematician

Papers and correspondence, incl mathematical texts and translations, notes and calculations 1651-95, papers on the Paschal Table and the calendar; autobiography; papers and correspondence rel to Oxford University, ciphers, and the Trinity; correspondence (*c*35 items) with the Revd T Smith 1677-99. *Bodleian Libr, Oxford*

Papers and correspondence, incl the first draft of his autobiography, extracts from mathematical works, letter book 1651-1701. *Brit Libr* (Sloane MSS 2284, 4025; Add MS 32499)

Treatise on logic; letters, incl 158 to H Oldenburg 1664-77, 18 to Sir H Sloane 1697-1700. *Royal Soc*

Papers on the Paschal Table, the calendar etc. *Christ Church, Oxford* (MS 165)

Problem propounded to B Frénicle de Bessy 1661; correspondence (*c*25 items) with C Huygens 1665-73. *Leiden Univ Libr*

Final draft of his autobiography. *Kent Archives Office, Maidstone* (U 120/F 15)

Proposal for the recoinage 1695. *London Univ Libr* (MS 62)

Paper (11ff) on the theory of music 1664. *Marsh's Libr, Dublin* (MS Z 3.4.24)

Correspondence with the Earl of Nottingham 1689-93. *Leicestershire RO*

Letters (44) to J Collins 1666-77. *Private*

▷CJ Scriba, 'A tentative index of the correspondence of John Wallis FRS', *Notes and Records*, xxii, 1967, pp58-93

[601] **WALSH, John** (1726-1795)
FRS, naturalist

Political and family correspondence and papers 1751-94. *India Office Libr and Records* (in Fowke MSS 20-39, MSS Eur D 546, G 37 and *passim*) [NRA 24532]

Diary of his journey to France 1772, incl notes of experiments on the electric ray. *John Rylands Univ Libr, Manchester* (Eng MS 724)

Notes and letters rel to his experiments on the electric ray in France 1772; paper on the gillaroo trout. *Royal Soc*

Deeds and papers rel to his estates in cos Cork and Kerry 1769-90. *Trinity Coll, Dublin* [NRA 21054]

Correspondence with Lord Clive. *Nat Libr of Wales*

[602] **WATERTON, Charles** (1782-1865)
Naturalist

Journal and other notes (4 vols) *c*1840-65. *Wakefield City Mus*

Deeds, estate and business papers and some correspondence 1807-65. *Wakefield District Archives* [NRA 23136]

Unpublished essay on living in the tropics 1838. *Brit Mus (Nat Hist) General Libr*

Correspondence with G Ord and others, and notes kept by his biographer Sir N Moore on visits to Waterton 1863-65. *Private*

Correspondence (36 items) 1841-65. *York City Archives* [NRA 9202]

Letters (13) to T Allis 1836-62. *Ibid* [NRA 9540]

[603] **WATSON, Hewett Cottrell** (1804-1881)
Botanist

Botanical papers, incl floras of various parts of Britain and other material on plant distribution for his *Cybele Britannica*, *Topographical botany* etc; letters (120) to Sir WJ Hooker 1830-52. *R Botanic Gardens, Kew*

Material for his *Cybele Britannica* and its supplements. *Brit Mus (Nat Hist) Botany Dept*

Letters (21) to Sir WC and Lady Trevelyan. *Newcastle Univ Libr* [NRA 12238]

Letters (11ff) to Sir W Jardine 1836-46. *R Scottish Mus, Edinburgh*

[604] **WATT, James** (1736-1819)
FRS, engineer

Correspondence and papers rel to his early activites in Scotland, incl journals, ledgers, notebook of experiments, reports, papers and plans concerning canals, harbours and rivers, and correspondence and papers rel to Boulton & Watt incl office and foundry letter books, engineering drawings (36,000) of steam engines and other machinery, engine books, ledgers and order books. *Birmingham Reference Libr* [NRA 14609]
Correspondence and papers 1759-1819, incl reports and surveys, specifications and lists of patents, pocket books and journals, letter books, legal papers and drawings. *Private* [NRA 22549]
Miscellaneous contracts and accounting papers of Boulton & Watt 1787-98. *Edinburgh Univ Libr* (in MSS La Add 2, 5-7)
Boulton & Watt drawings. *Birmingham City Mus*
Papers (5) submitted to the Royal Society. *Royal Soc*

See also Black, Robison, Wedgwood J, Withering

[605] **WEBSTER, Thomas** (1772-1844)
Geologist

Autobiography 1837; correspondence and notes (8 items) 1800-04. *Royal Inst* (MS 121 A-B)
MS of article 'On the freshwater formations in the Isle of Wight' 1814 and drawings of fossils. *Geological Soc of London*
Student notes on his lectures 1827; some correspondence. *University Coll London* (MS Add 49)
Correspondence (*c*150 items) 1818-44. *University Coll of Wales, Aberystwyth*
Miscellaneous correspondence (51 items) 1814-38. *Fitzwilliam Mus, Cambridge* (SG Perceval Colln)

See also Buckland

[606] **WEDGWOOD, Josiah** (1730-1795)
FRS, potter

Correspondence and papers. *Keele Univ Libr*
Copies of correspondence (10 vols) 1758-95. *John Rylands Univ Libr, Manchester* (Eng MSS 1101-10)
Extracts made by him from scientific publications (10 vols). *Brit Libr* (Add MSS 28309-18)
Papers (6) submitted to the Royal Society. *Royal Soc*
Correspondence (65 items) with J Watt 1779-89. *Private* [NRA 22549]
See also Priestley

[607] **WEDGWOOD, Thomas** (1771-1805)
Pioneer of photography

Correspondence and papers. *Keele Univ Libr*
Papers (2) submitted to the Royal Society. *Royal Soc*
Copies of letters addressed to him 1799-1804. *John Rylands Univ Libr, Manchester* (Eng MS 1110)

[608] **WESTWOOD, John Obadiah** (1805-1893)
Entomologist

Entomological correspondence and papers, incl diaries and drawings. *Hope Libr, University Mus, Oxford*
Entomological drawings (3 vols) *c*1832, and drawings (75) of insects from Madeira and the Canary and Salvage Is 1857-60. *Brit Mus (Nat Hist) Entomology Dept*
Correspondence, articles, catalogues of insects and miscellaneous papers 1816-90. *Smithsonian Inst, Washington* (MS 7112)
Correspondence (103ff) 1842-86. *Bodleian Libr, Oxford* (in MS Autogr d 21)
Letters (17) to W Swainson 1830-40. *Linnean Soc*
Drawings (12) of Indian insects. *India Office Libr and Records* (in NHD 5)

[609] **WHEATSTONE, Sir Charles** (1802-1875)
FRS, physicist

Correspondence and papers, incl notes on light, aniline dyes etc, and lectures on sound. *King's Coll, London*
MSS of papers published in *Philosophical Transactions*; correspondence (26 items) with Sir JFW Herschel 1825-49. *Royal Soc*
Papers rel to the arbitration between Wheatstone and Sir WF Cooke. *Inst of Electrical Engineers* [NRA 20573]
Drawings (*c*100), mainly rel to the development of the electric telegraph, 1837-65. *Science Mus Libr*
Letters (11) to C Babbage 1839-43. *Brit Libr* (in Add MSS 37191-92, 37201)

[610] **WHELER, Revd Granville** (1701-1770)
FRS, natural philosopher

Papers on electricity 1737-38. *Brit Libr* (in Add MSS 4433-34)
Paper submitted to the Royal Society. *Royal Soc*

[611] **WHEWELL, Revd William** (1794-1866)
FRS, man of science

Correspondence and papers, incl *c*7,000 letters addressed to him, journals and notebooks 1817-53, geological sketch book

1828, MSS of his lectures and publications, notes on geology, mineralogy, mathematics etc. *Trinity Coll, Cambridge* [NRA 8804]
Correspondence (*c*175 items) with Sir JFW Herschel 1817-66; letters (87) to Sir JW Lubbock 1823-47. *Royal Soc*
Letters (139) to JD Forbes 1831-65. *St Andrews Univ Libr* [NRA 13132]
Correspondence (126 items) with Sir GB Airy 1841-64. *R Greenwich Observatory, Herstmonceux* (in MSS 939-46, 948-51)
Letters (37) to Sir RI Murchison. *Geological Soc of London*
Letters (34) to A Quételet 1832-61. *Bibl Royale, Brussels*
Letters (28) to various correspondents 1828-63. *Science Mus Libr* (in MSS 1098-1125)
Letters (26) to Lord Monteagle 1850-54. *Nat Libr of Ireland* (MS 13401)
Correspondence (25ff) with the Revd C Wordsworth 1828-53. *Lambeth Palace Libr* (in MSS 2140-44)
Letters (15) to Sir JFW Herschel 1832-58. *Texas Univ, Austin*
Letters (13) to Mary and W Somerville 1831-45. *Bodleian Libr, Oxford* [NRA 9423]
Letters (10) to C Babbage 1820-47. *Brit Libr* (Add MSS 37182-37201 *passim*)
Miscellaneous correspondence (17 items) 1838-56. *Houghton Libr, Harvard Univ, Cambridge, Mass*

▷ RP Sturges, *Economists' papers 1750-1950: a guide to archive and other manuscript sources for the history of British and Irish economic thought*, 1975, pp126-28

See also Faraday, Herschel JFW, Lyell, Murchison, Sedgwick, Sheepshanks, Somerville

[612] **WHITE, Revd Gilbert** (1720-1793)
Naturalist

Papers, incl original MS with additions and annotated family copy of the *Natural history and antiquities of Selborne*, sermons (6), treatise on man's social obligations, brewing journal 1772-93, personal, household, estate and annuity accounts 1745-93, and family papers. *Gilbert White Mus, Selborne* [NRA 11555]
'Garden-Kalendar' 1751-67; 'The Naturalist's Journal' (6 vols) 1768-93; annotated copy of T Pennant's *British zoology*, 2nd edn (1768); letters (30) to T Pennant 1767-73; letters (14) to D Barrington 1769-80. *Brit Libr* (Add MSS 31846-52, 35138-39)
Papers 1739-93, incl account books, 10 letters from R Churton 1786-93, and family papers. *Houghton Libr, Harvard Univ, Cambridge, Mass* (MS Eng 731) [NRA 20129]
Account book 1739-46 and sermons 1762-92. *Oriel Coll, Oxford* (Case D C III.11-12)
Sermon on repentance (1 vol), incl list of 31 dates and locations of preaching it 1757-92; MS of his 'Advertisement' to the *Natural history* 1788. *Private*
MS of his translation of Cicero's *De oratore*, III, 1741. *Private*
Papers (3) submitted to the Royal Society. *Royal Soc*
Letters (233) from the Revd J Mulso 1744-90; letters (35) to his family 1758-93. *John Rylands Univ Libr, Manchester* (Eng MSS 911, 1306)
Letters (60) to Mary White 1778-91. *Fitzwilliam Mus, Cambridge*
Letters (17) to R Churton 1779-93. *Bodleian Libr, Oxford* (MS Eng misc b 9)

[613] **WHITTAKER, Sir Edmund Taylor** (1873-1956)
FRS, mathematician

Papers, incl referee reports 1905; letters (8) to Sir J Larmor; miscellaneous letters (4) 1915-20. *Royal Soc*
Correspondence with his son JM Whittaker 1924-56. *Private*
Correspondence with O Veblen 1930-38. *Libr of Congress, Washington*
Correspondence (37 items) with H Dingle 1938-56. *Imperial Coll, London* [NRA 24527]
Correspondence with Viscount Cherwell 1949-50. *Nuffield Coll, Oxford* [NRA 16447]
Letters to officers of the Royal Astronomical Society from 1898. *R Astronomical Soc*

[614] **WHITWORTH, Sir Joseph,** 1st Bt (1803-1887)
FRS, engineer

Letters (17) to C Babbage 1847-69. *Brit Libr* (in Add MSS 37193-97, 37199)
Correspondence (11 items) with Sir E Chadwick 1842-75. *University Coll London* [NRA 21653]

[615] **WILLIAMS, Arthur Stanley** (1861-1938)
Astronomer

Correspondence 1880-1938 and papers, incl observation journals (51 vols) 1877-1938, papers rel to variable stars (62 files), Jupiter (35 files) and Saturn (6 files), miscellaneous notes (8 vols), letter books (6 vols) and 90 letters to officers of the Royal Astronomical Society 1880-1900. *R Astronomical Soc*

[616] **WILLIS, Revd Robert** (1800-1875)
FRS, mechanician

Lectures and papers on mechanics. *Cambridge Univ Libr* (MSS Add 5131, 5133-34, 5136)

Letters (15) to the Society for the Diffusion of Useful Knowledge 1834-43, with copies of his correspondence about British machinery 1841-42. *University Coll London* [NRA 22241]

[617] **WILLIS, Thomas** (1621-1675)
FRS, physician

'Pathologia cerebri et de scorbuto' 1667; 'Pharmaceutice rationalis' 1674. *Brit Libr* (in Sloane MS 1845)
Copies of prescriptions by Sir TT de Mayerne. *R Coll of Physicians, London*

[618] **WILLOUGHBY, Francis** (1635-1672)
FRS, naturalist

Papers, incl commonplace book with notes on architecture, philosophical, moral and religious topics, classical literature, geography, geometry, chemistry, botany and medicine 1658-65; English-Welsh vocabulary 1662; rolls of vocabularies of many European and Asian languages compiled during his travels in Europe (1663-66); MS account corrected by him of a dissection at Padua and related discourses and lectures 1663-64; volume of notes on games; library catalogue; volume of extracts rel to Willoughby family history begun by him and continued by the Revd J Ray; drawings. *Nottingham Univ Libr* (Middleton MSS)
'Piscium synopsis' and 'Ornithologiae synopsis'. *Brit Libr* (in Sloane MSS 411, 618)
Papers (3) submitted to the Royal Society; letters (16) to H Oldenburg 1667-71. *Royal Soc*

▷MA Welch, 'Francis Willoughby, FRS (1635-1672)', *J Soc Bibliog Nat Hist*, vi, pt 2, 1972, pp71-85

See also Ray

[619] **WILSON, Charles Thomson Rees** (1869-1959)
FRS, physicist

Laboratory notebooks (*c*40); referee reports (8) 1896-1910. *Royal Soc*
Papers (2 boxes) on the effects of thunderstorms on balloons and airships 1918-40. *Churchill Coll, Cambridge*

[620] **WILSON, William** (1799-1871)
Botanist

Drawings and notes for *Bryologia Britannica* (1885); correspondence rel to mosses (12 vols). *Brit Mus (Nat Hist) Botany Dept*
MS of *Bryologia Britannica* 1854, related notebooks and papers 1827-67, and botanical correspondence (4 vols). *Warrington Public Libr* [NRA 9201]
Correspondence (678ff) 1827-69. *New York Botanical Garden Libr*
List of mosses in Sir WJ Hooker's herbarium 1840, and 200 letters to Hooker 1827-65. *R Botanic Gardens, Kew*

See also Thwaites

[621] **WILSON, William Edward** (1851-1908)
FRS, astronomer and physician

Letters (42) to officers of the Royal Astronomical Society 1875-1900. *R Astronomical Soc*
Correspondence with GE Hale 1904-06. *California Inst of Technology, Pasadena*

[622] **WINCH, Nathaniel John** (1768-1838)
Botanist

Letters (1,413) addressed to him, and annotated copies of botanical works by himself and others. *Linnean Soc*
Letters (62) to Sir WC and Lady Trevelyan 1822-35. *Newcastle Univ Libr* [NRA 12238]
Letters (31) to Sir WJ Hooker 1830-38. *R Botanic Gardens, Kew*

[623] **WITHERING, William** (1741-1799)
FRS, physician

Working papers rel to fungi, hydrocephalus etc. *R Coll of Surgeons, London*
Papers submitted to the Royal Society on toadstone and Rowley rag, heavy spar, lightning, and J Priestley's principle of acidity. *Royal Soc*
Drawings of British plants (540). *Wellcome Hist Medical Libr* (MS 5035)
Correspondence 1781-99, together with copies of correspondence 1775-99 and other biographical material. *Birmingham Univ Libr* [NRA 13196]
Letters (34) to J Watt 1780-99. *Private* [NRA 22549]
Some correspondence. *Royal Soc of Medicine, London*

[624] **WOLLASTON, William Hyde** (1766-1828)
FRS, man of science

Notebooks (19), and correspondence. *Cambridge Univ Libr* (MS Add 7736)
Notebook on platinum and rhodium 1822-25. *Science Mus Libr*
Papers (30) submitted to the Royal Society 1797-1805 and nd; diplomas; correspondence (15 items) with Sir JFW Herschel 1823-27; letters (9) to T Young 1800-01. *Royal Soc*

Copies of letters (60) to the Revd H Hasted 1797-1828 (originals lost). *University Coll London* (Gilbert Papers)

[625] **WOODWARD, Sir Arthur Smith** (1864-1944)
FRS, palaeontologist

Correspondence and papers, incl material rel to the Piltdown skull and autobiographical notes. *Brit Mus (Nat Hist) Palaeontology Dept*

Correspondence 1880-1939. *University Coll London* (MS Add 242)

Letters (71ff) to Macmillan & Co 1924-32. *Brit Libr* (Add MS 55224)

See also Seeley

[626] **WOODWARD, John** (1665-1728)
FRS, geologist and physician

Correspondence and papers rel to geology, medicine etc *c*1695-1727. *Cambridge Univ Libr* (MS Add 7647)

MS (6 vols) of his *Attempt towards a natural history of the fossils of England c*1725. *Sedgwick Mus of Geology, Cambridge*

Discourses on metals. *Brit Libr* (Add MSS 25095-96)

Letters (*c*80) to T Hearne 1707-26. *Bodleian Libr, Oxford* (in MS Rawl lett 12)

Letters (34) to Sir H Sloane 1698-1723. *Brit Libr* (Sloane MSS 4037-62 *passim)*

Letters (30) to E Lhuyd *c*1692-1694. *Bodleian Libr, Oxford* (in MS Ashmole 1817b; MS Eng hist c 11)

Letters (12) from J Hutchinson 1706; letters (12) from the Revd B Holloway 1719-28. *Ibid* (in MS Gough Wales 8)

▷ VA Eyles, 'John Woodward, FRS, FRCP, MD (1665-1728): a bio-bibliographical account of his life and work', *J Soc Bibliog Nat Hist*, v, pt 6, 1971, pp399-427

[627] **WOODWARD, Samuel Pickworth** (1821-1865)
Naturalist

Nature notes made at Norwich 1832-34. *Brit Mus (Nat Hist) General Libr*

Notes on Geological Society discussions of papers by W Buckland and Sir C Lyell on glaciers 1840. *Inst of Geological Sciences* [NRA 18675]

Illustrated catalogue of fossils in the cabinet of Mrs MH Smith 1845; drawings of brachiopoda. *Brit Mus (Nat Hist) Palaeontology Dept*

MS of and drawings for his *Manual of the mollusca* (1851-56), and other notes and drawings on mollusca. *Ibid, Zoology Dept*

Notes and extracts (1 vol), and annotated working copy of his *Manual of the mollusca* (1851-56). *Linnean Soc*

Some correspondence. *Castle Mus, Norwich*

See also Turner D

[628] **WREN, Sir Christopher** (1632-1723)
FRS, natural philosopher and architect

Drawings and plans (*c*400); transcripts (1 vol) rel to his life and works in mathematics and architecture 1728. *All Souls Coll, Oxford*

Accounts and other papers 1663-1723 rel to his work on St Paul's Cathedral and the rebuilding of London parish churches. *Guildhall Libr, London*

Memoranda, accounts and agreements (1 vol) rel to Winchester Palace *c*1682-84; note of Greenwich fees and payments 1669-73. *Royal Inst of British Architects* [NRA 13990]

His report on his designs for the Monument 1675. *Brit Libr* (Add MS 18898)

Account book as surveyor of the king's works 1682-83. *Nottinghamshire RO* (Foljambe of Osberton Papers) [NRA 20442]

Proposal for the recoinage 1696. *London Univ Libr* (MS 62)

Reports on lands for churches in Cripplegate 1712 and Lower Wapping 1715. *Lambeth Palace Libr* (in MS 2714)

Description of a level invented by him and produced before the Royal Society, with sketch, nd. *Brit Libr* (in Sloane MS 3323)

'Parentalia': papers of Wren and other biographical material collected by his son. *Royal Soc* (MS 249)

[629] **WRIGHT, Sir Almroth Edward,** KBE (1861-1947)
FRS, bacteriologist

Log annotated by Wright of typhoid experiments carried out on him and others at the Army Medical School, Netley, 1901. *Brit Libr* (Add MS 56138)

Correspondence and papers mainly rel to his own research. *Medical Research Council*

Correspondence with GE Hale 1931-32. *California Inst of Technology, Pasadena*

Letters (9) to GB Shaw 1906-43. *Brit Libr* (in Add MS 50553)

[630] **WRIGHT, Edward** (?1558-1615)
Mathematician and hydrographer

Astronomical collections *c*1600; observations. *Trinity Coll, Dublin* (MSS 387, 396)

'The description and use of the celestial automaton'; 'Examples in the art of fortification, with diagrams'. *Brit Libr* (Sloane MS 651)

[631] **WRIGHT, Thomas** (1711-1786)
Natural philosopher

Papers (8 vols). *Tyne and Wear Archives Dept, Newcastle upon Tyne*
Autobiography down to 1762, and illustrated observations on the antiquities of England and Ireland. *Brit Libr* (Add MSS 15627-28, 33771)
Papers, incl 'The universal vicissitude' 1737, 'Pansophia or an essay towards a general compendium of universal knowledge' *c*1750, 'An essay towards a complete chronological catalogue of philosophers, astronomers and mathematicians', and notes for a second edition of his *Original theory or new hypothesis of the universe* (1750). *Durham Univ Libr* (MSS Add 168, 322 etc)
Papers read to the Royal Society on astronomy 1735-74 and on the ancient earth works at Buckland Rings, Hampshire 1745. *Royal Soc*

[632] **YOUNG, Arthur** (1741-1820)
FRS, agriculturist

Original MS (34 vols) and transcript of his unpublished treatise 'The elements and practice of agriculture'; correspondence (8 vols) 1743-1820. *Brit Libr* (in Add MSS 34821-64, 35126-33)
Autobiographical account (11ff) in form of letters to Jane Young 1809-10, and some family correspondence. *John Rylands Univ Libr, Manchester* (Bagshawe MSS)
Minutes kept by him as secretary of the Board of Agriculture and related correspondence and papers. *Inst of Agricultural Hist, Reading Univ* (Royal Agricultural Society of England Records) [NRA 20986]
Letters (*c*100) to Marianne Francis, Charlotte Broome and Charlotte Barrett 1811-20. *New York Public Libr* (Berg Colln)
Letters (26) to Sir J Banks 1786-99. *Sterling Memorial Libr, Yale Univ, New Haven*
Letters (26) to the Earl of Egremont 1793-1815. *Private*
Letters (12) to T Ruggles 1795-1811. *Private*
Letters (10) to the Earl of Hardwicke 1799-1819. *Brit Libr* (in Add MSS 35643, 35652, 35697, 35700)
Miscellaneous letters (13) 1770-1802. *R Soc of Arts*

▷ JG Gazley, 'Arthur Young, agriculturist and traveller, 1741-1820. Some biographical sources', *Bulletin of the John Rylands Library Manchester*, xxxvii, no 2, 1955, pp393-428; RP Sturges, *Economists' papers 1750-1950: a guide to archive and other manuscript sources for the history of British and Irish economic thought*, 1975, pp131-32

[633] **YOUNG, James** (1811-1883)
FRS, chemist and fuel technologist

Correspondence and papers, incl notebooks and diaries 1831-83, letter books 1860-73, account book 1863-67, and miscellaneous legal papers. *Andersonian Libr, Strathclyde Univ, Glasgow* [NRA 9394]
Letters (36) from D Livingstone 1858-72. *Nat Mus, Livingstone, Zambia*

See also Playfair

[634] **YOUNG, Revd Matthew** (1750-1800)
Natural philosopher

Papers (4 vols), chiefly mathematical. *Trinity Coll, Dublin* (MS 949)

[635] **YOUNG, Thomas** (1773-1829)
FRS, physician and physicist

Correspondence and papers, incl 52 letters to him; correspondence (41 items) with Sir JFW Herschel 1821-29. *Royal Soc*
Notes (20 vols) for his lectures at the Royal Institution 1802-03. *University Coll London* (MS Add 13)
Papers rel to his Egyptological studies, plates for his *Hieroglyphics* (1823-28), and correspondence on hieroglyphic literature 1814-28. *Brit Libr* (Add MSS 21026-27, 27281-85)

See also Wollaston

Index of locations

Institutions for which no location is indicated in the text are situated in London

Printed in England for Her Majesty's Stationery Office by Albert Gait Ltd., Grimsby
Dd. 696396 C.30